W0246433

Systems Theory and Application

This book describes the characteristics of five different disciplines of systems which are Systems Theory, Systems Science, Thinking in Systems, Systems Architecting, and Systems Engineering. The book discusses how they all relate to each other and form a synergistic set of disciplines.

Systems Theory and Application: A Multi-Disciplinary Approach presents how the five different disciplines of systems are all related to each other. The book offers a concise view of the systems perspective and discusses how it applies to many system types such as physical, abstract, and human. Highlights are on how systems disciplines address problems and abandon the fragmented approach of implementing the disciplines separately. The book forms an enlightenment on understanding the relationship between systems engineering and system theory and explains that systems are everywhere, and that universe is made up of systems.

Students, designers, and those interested in systems theory will find this book of interest.

Systems Theory
and Application
A Multi-Disciplinary Approach

Scott Jackson
Ricardo Moraes

CRC Press
Taylor & Francis Group
Boca Raton London New York

CRC Press is an imprint of the
Taylor & Francis Group, an **informa** business

First edition published 2024
by CRC Press
2385 NW Executive Center Drive, Suite 320, Boca Raton FL 33431

and by CRC Press
4 Park Square, Milton Park, Abingdon, Oxon, OX14 4RN

CRC Press is an imprint of Taylor & Francis Group, LLC

ISBN: 978-1-032-58085-2 (hbk)
ISBN: 978-1-032-58408-9 (pbk)
ISBN: 978-1-003-44993-5 (ebk)

DOI: 10.1201/9781003449935

Typeset in Times New Roman
by SPi Technologies India Pvt Ltd (Straive)

Contents

About the Authors

Scott Jackson is a Fellow of the International Council on Systems Engineering (INCOSE). He has written four books on systems engineering. He is a consultant for both Embraer of Brazil and Comac of China and was formerly with the University of Southern California (USC) and the Missouri University of Science and Technology (MST).

Ricardo Moraes is currently an Engineering Manager at Embraer, with experience in solution development and a focus on solution architectures and complex systems in sectors such as commercial aviation, executive aviation, and defense programs. He is also the president of the INCOSE Brazil Chapter.

Systems Theory

1

In this book the term "Systems Theory" encompasses all the disciplines discussed. This idea is illustrated in Figure 1.1.

This figure shows how the various systems disciplines are related and are discussed in this book. The primary reference for Systems Theory is [1]. It can be seen in Figure 1.1 that Systems Theory encompasses Systems Science, Systems Engineering, Systems Thinking, and Systems Architecting. So, what do all these disciplines have in common? The definition of a system for one thing.

1.1 **Systems Science**. The term "Systems Science" applies to all of the phenomena that occur in nature and in human-made systems that can be explained from a systems perspective. Other disciplines, such as Systems Engineering, also employ the principles of Systems Science. The primary reference for Systems Science is [2].

1.2 **Systems Engineering**. This is the practical arm of Systems Theory. A primary reference for Systems Engineering is [3]. The Application arrow is shown from Systems Theory.

1.3 **Systems Architecting**. The purpose of Systems Architecting is to determine the arrangement of the parts of the system before it is subjected to the processes of Systems Engineering. A primary reference for Systems Architecting is [4].

1.4 **Systems Thinking**. According to [5], Systems Thinking "is about a particular way of thinking about the world, one which although broadly a part of the science movement, uses some concepts which are complementary to those of classical natural science." Referring back to Figure 1.1, Systems Thinking enables all the disciplines in that diagram. Prominent books on this subject include [5, 6].

DOI: 10.1201/9781003449935-1

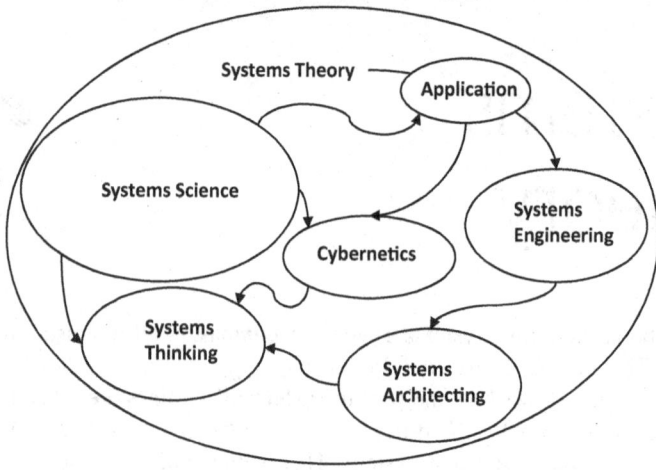

FIGURE 1.1   System disciplines.

SYSTEM

Although many definitions of the term "system" exist, a convenient one to use is the one posed by the International Council on Systems Engineering (INCOSE) Fellows in 2018 [7]. The essential aspects of a system are as follows:

- A system must consist of parts.
- The parts must interact.
- The collection of parts must exhibit a property called emergence which cannot be identified by the examination of a single part.
- The system must have a boundary. This aspect will be discussed later.
- The system may have other characteristics listed by the INCOSE Fellows in 2018 [7]. We have listed these characteristics in Appendix C.

Types of Systems

Systems can be divided into two major categories: natural and human-made. The purpose of this chapter is to identify as many system types and examples as possible and to restate the basic theme that they are all systems as defined in this section and that they all interact in many ways.

Natural systems

The main characteristic of natural systems is that they occur naturally without any help from humans even though humans often attempt to modify them, augment them, and even copy them. Common natural systems include the solar system, the human body, trees, and humanity.

One person particularly interested in natural systems is Professor Neri Oxman of MIT. Professor Oxman has produced a major collection of natural systems [8] whose characteristics have been described. According to the publisher, Allianz, Oxman has ambitious goals:

> Oxman envisions a future in which these disciplines work together to directly address the existential challenge of our time: the securing of a sustainable future for us all.

Hence, Oxman's vision is more than the depiction of a system, but rather the multidisciplinary vision of the future. Oxman goes on to say that her approach is a "method of design and production that brings together humans, automated processes, and nature to transform architecture into a hybrid act of building and growing."

She calls her field Material Ecology which she defines as

> the study of design of products and processes integrating environmentally aware computational form-generation and digital fabrication, the field operates at the intersection of biology, material science and engineering, and computer science, with emphasis on environmentally informed digital design and fabrication.

Her emphasis is System over Object which she defines as "the product—be it a product-a wearable device, or a building—is considered part of a system of interrelations between natural and designed environments including interactions between the entity and the human body between the entity and the environment." As can be seen, she maintains the system emphasis with a broad scope.

The primary emergent property of natural systems is their ability to exist and grow without the aid of humans.

The solar system

It would be difficult to categorize the solar system as anything except natural. There are no current plans to add or remove planets. However, NASA is studying the possibility of nudging one or more asteroids to keep them from hitting the Earth.

All the planets could be considered elements of the solar system and other objects such as asteroids and comets as well. One emergent property of the solar system is the ability to travel in orbits around one another.

The human body

The human body has long been recognized as a system consisting of 11 major subsystems. The human body is a natural system. All of these subsystems are studied in medical schools. They are the circulatory system and ten others. It is recognized that these subsystems are not totally independent and interact with each other. A good source for a description of the human system is [9]. There are many examples of emergent properties of the human system. The ability to give birth to another human is one of them. This book presents the human system in several ways:

- Chapter 4 presents the human as an essential element as part of the restructuring resilience principle as in the Apollo 13 mission. In this example, a human would need to know how a system could be restructured to achieve system recovery.
- Also in Chapter 4 it can be seen that a human was required to land the spacecraft on the moon when it was determined that an auto-mated landing might not be possible.
- Finally, Chapter 8 explores how artificial intelligence (AI) might be able to mimic human actions to make systems safer.

Trees

Trees are good examples of natural systems as are flowers and other plants. Trees are a good opportunity to introduce the idea of a *holon*. Holons are parts of other systems that contain entire systems within themselves. A seed is a good example of a holon. Each seed contains other seeds and other systems. Theoretically there is an infinite number of holons in a seed. The emergent property of trees is their ability to produce holons of themselves.

Humanity

Humanity is also a natural system although many people are reluctant to make that claim. Subsystems come in many forms, the most obvious being the human body as described above. Other subsystems consist of ethnic groups. The emergent property of humanity is its ability to create civilizations.

Organizational

Although organizational systems consist of many natural systems, the organization itself is a human-made system. Typical organizational systems include companies and armies. The emergent property of organizational systems is their ability to provide services and products.

Physical Systems

Physical systems are, in general, human-made.

Aircraft

Aircraft, in particular, commercial aircraft, are human-made systems and include the pilot or operator. When the pilot is removed from the aircraft, the remaining elements are called artifacts. When an aircraft is defined and integrated, two processes for doing this are systems architecting and systems engineering. The book *Systems Engineering for Commercial Aircraft: A Domain Specific Adaptation* [10] describes these processes. The primary emergent properties of a commercial aircraft are its purpose, its name, and its ability to transport people and cargo over a long distance.

Abstract Systems

Abstract systems exist only in the minds of their creators. Abstract systems are human-made. This definition leads to the conclusion that this book is an abstract system. Within the Systems community, it is a common practice to refer to the study of systems as a *meta-system*. This is because this meta-system is beyond all normally recognized systems, whose purpose is to show how all the other systems relate to each other.

Plans

The plans to build a commercial aircraft are an example of an abstract system. However, the abstract system exists only in the mind of the creator. When it is put on paper or entered into a digital database, it is no longer abstract. The emergent property of a plan is its ability to instruct the creator on how to create the physical system from it.

Laws

Laws are abstract systems intended to provide instructions to the public on what is permissible or not permissible.

Symphonies

Symphonies and other forms of music are abstract systems designed to instruct musicians on how to play the intended score. The emergent property of a system is its organization including its motif and melody.

Government

A government system is a common abstract system. It consists of the governmental organization and the laws that put it into pace. An emergent property of an abstract government system is the physical manifestation of that system.

Abstracted System

Another system type identified by [11] is the abstracted system This type of system occurs when a pure abstract system becomes part of a real physical system.

Complex Systems

Complex systems are a type of system that is found abundantly in the real world. According to Watson et al. [12, 13], "complex systems tend to exhibit variety in their characteristics rather than uniform repetition." Watson et al. provide a set of 14 distinguishing characteristics that are commonly found in complex systems. These characteristics are summarized as follows:

Diversity: The structural, behavior, and system state varieties that characterize a system and/or its environments.

Connectivity: The connection of the system between its functions and the environment. This connectivity is characterized by the number of nodes, diversity of node types, number of links, and diversity in link characteristics. Complex systems have multiple layers of connections within the system structure. Discontinuities (breaks in a pattern of connectivity at one or more layers) are often indications of complex system connectivity. Simple and some complicated systems may be characterized by simpler structures such as hierarchies.

Interactivity: The behavior stimulus and response between different parts of a system and the system with its environment. Complex systems have many diverse sources of stimulus and diverse types of responses. The correlation between stimulus and response can be both direct and indirect (perhaps separated by many layers of system connectivity). The types of stimuli and responses vary greatly. The levels of stimuli and responses can range from being very subtle to very pronounced. The timeframe for system responses can vary hugely.

Adaptability: Complex systems proactively and/or reactively change their function, relationships, and behavior to balance the changes in the environment and to achieve system goals.

Multiscale: Behavior, Relationships, and Structure exist on many scales, are ambiguously coupled across multiple scales, and are not reducible to only one level.

Multiperspective: Multiple perspectives, some of which are orthogonal, are required to comprehend the complex system.

Behavior: Complex system behavior cannot be described fully as a response system. Complex system behavior includes nonlinearities. Optimizing system behavior cannot often be done by focusing on properties solely within the system.

Dynamics: Complex systems may have equilibrium states or may have no equilibrium states. Complex system dynamics have multiple scales or loops. Complex systems can stay within the dynamical system or generate new system states or state transitions due to internal system changes, external environment changes, or both. Correlation of the changes in complex systems to events or conditions in the system dynamics may be ambiguous.

Representation: Representations of complex systems can be difficult to properly construct with any depth. It is often impossible to predict future configurations, structures, or behaviors of a complex system, given finite resources. Causal and influence networks create a challenge in developing "requisite" conceptual models within these time and information resource constraints.

Evolution: Changes over time in complex system states and structures (physical and behavioral) can result from various causes. Complex system states and structures are likely to change as a result of interactions within the complex system, with the environment, or in application. A complex system can have disequilibrium (i.e., non-steady) states and continue to function. Complex system states and structures can change in an unplanned manner and can be difficult to discern as they occur. The changes in the states and structure of a complex system are a natural function of (is often present in) the complex system dynamics. Changes can occur without centralized control, due to localized responses to external and/or internal influences.

Unexpected Emergence (Complex): Emergent properties of the holistic system are unexpected (whether predictable or unpredictable) in the system functionality/response. They are unpredictable given finite resources. Behavior is not describable as a response system.

Disproportionate Effects: Details seen at fine scales can influence large-scale behavior. Small-scale modifications can result in radical changes of behavior. Scale can be in terms of magnitude of effect or an aggregate amount of change. Weak ties can have disproportionate effects.

Indeterminate boundaries: Complex system boundaries are intricately woven with their environment and other interacting systems. Their boundaries can be non-deterministic. The boundary cannot be distinguished based solely on processes inside the system.

Contextual Influences: All systems reside in natural and social environments and relate to these. In the relationship between the system and the natural and social environments, there can be complexity. This complex interaction depends on the social application of the system. Social systems often strive to achieve multiple, sometimes incompatible, objectives with the application of the same system.

Complicated Systems

In the world of systems, complicated systems are different from complex systems in that, according to Watson et al. [13], complex systems involve the interaction of dynamic parts, while complicated systems involve the interaction of static parts.

Systems of Systems

This section shows how several systems can be linked together and called a system of systems (SoS). It also shows how this combination of systems can be called a system and how that system can have emergent properties.

Watson et al. [13] provide a set of rules for managing complex systems:

- Complex systems need balance rather than optimization.
  The whole is often sub-optimized when a part is optimized, or an optimized system can become rigid and cannot cope with changing circumstances and needs.
- Tension is common in complex systems. Tension between large and small, distributed and central, agile and planned, calls for perpetual seeking of balance.

- Complexity can be bounded within a simpler structure. For example, biological cells are internally complex and yet a single cell is a simple structure externally.
- The characterization of a system architecture in terms of some coordinated collection of subsidiary design elements (e.g., responsible "trades" for specific design elements) can be a major step toward organizing and managing complexity. Implicit in successfully navigating this task is mutual appreciation between responsible trades regarding their respective contributions and interactions to the evolving system(s).
- Social systems often strive to achieve multiple, sometimes incompatible, objectives engendering risks and opportunities for the coevolution of systems services and value to society. Better understanding of these co-evolutionary processes may prove useful to engineering organizations.

Assignment: Identify a complex system and explain why it is not complicated.

REFERENCES

[1] L. V. Bertalanffy, *General Systems Theory: Foundation, Development, Applications*, Revised ed. New York: George Baziller, 1968.
[2] G. Mobus and Michael C. Kalton, *Principles of Systems Science (Systems Science)*. New York: Springer, 2015.
[3] B. Blanchard and W. J. Fabrycky, *Systems Engineering and Analysis*, 4th ed. (Prentice Hall International Series in Industrial and Systems Engineering). Upper Saddle River, NJ: Prentice Hall, 2006. 1981.
[4] M. W. Maier and E. Rechtin, *The Art of Systems Architecting*, 3rd ed. Boca Raton, FL: CRC Press, 2009. 1991.
[5] P. Checkland, *Systems Thinking, Systems Practice*. New York: John Wiley & Sons, 1999.
[6] D. H. Meadows and D. Wrighr, Ed. *Thinking in Systems*. White River Junction, Vermont: Chelsea Green Publishing, 2008.
[7] H. G. Sillitto et al., "Defining 'System': A Comprehensive Approach," presented at the IS 17, Adelaide, Australia, July 15–20, 2017.
[8] N. Oxman, *Material Ecology*. New York: Museum of Modern Art, 2020.

[9] Britannica. "The Human Body." Encyclopedia Brittanica. https://www.britannica.com/science/human-body (accessed March 3, 2023).

[10] S. Jackson and G. Loft, Eds. *Systems Engineering for Commercial Aircraft: A Domain Specific Adaptation*, 2nd ed. Aldershot, UK: Ashgate Publishing Limited (in English and Chinese), 2015.

[11] H. G. Sillitto et al. "System and SE Definitions." International Council on Systems Engineering. https://www.definitions.sillittoenterprises.com/ (accessed August 8, 2018).

[12] M. D. Watson. (2019, May 18–28) Systems Engineering Principles and Hypotheses. *Insight.*

[13] M. Watson, R. Anway, D. McKinney, L. A. Rosser, and J. MacCarthy, "Appreciative Methods Applied to the Assessment of Complex Systems," presented at the *International Council on Systems Engineering (INCOSE) International Symposium*, Orlando, FL, 2019.

Systems Science

2

A primary source of systems science is Mobus [1]. The principles of systems science are generally at a higher level of abstraction. Therefore, their use in associated disciplines must be derived.

Mobus [1] defines systems science as

> a science that crosses borders to address whole relational systems would be a metascience, "meta" being Greek for "beyond". When we call systems science, the idea is not that systems science is literally beyond science, but it deliberately goes beyond the boundaries of any particular science to include them all. Such systems science would not know physics are chemistry better than physicists or chemists, but since it studies the complex systemic relationships that are broken down into diverse fields of study, systems science should have more to say than physicists and chemists regarding the relationship of physics to chemistry or both to biology, etc.

The reader now having learned what the term "meta" means will now understand why this book is a meta-system. Hence, it can be said that systems science does not intend to deny the right of physicists and chemists to describe phenomena in their own fields; however, it has the goal of describing those phenomena in a systems context. It shows that it is that context itself that is the root cause of the various phenomena. Emergent properties, to be described later, is a good example.

EMERGENCE

One of the most striking examples of a contribution of systems science to systems theory is the phenomenon, or rather the various phenomena called "collective emergence." Emergence comes in various forms and is stimulated in

DOI: 10.1201/9781003449935-2

FIGURE 2.1   Clash of philosophies.

different ways, so it is not always easy to understand. This section will try to make it easier.

According to Checkland [2], emergence is "the principle that whole entities exhibit properties which are meaningful only when attributed to the whole not to its parts." Checkland lists the smell of ammonia as an example.

Before we leave Checkland's example behind, let's examine it a bit. First of all, ammonia is a compound consisting of both odorless nitrogen and hydrogen gases. So, why does ammonia have an odor? The simple answer is that it is the emergent reaction to the two gases.

What else can we conclude? The answer is that emergence only occurs in a systems context. Since there are two gases present, that condition is met. The second conclusion is that non-systems do not exhibit emergence. For example, nitrogen alone is a non-system; hence, there is no odor, that is, no emergent property.

Salt

Another common example is salt, which is common table salt that consists of sodium and chlorine. The result is the same as for ammonia: without emergence, the salt would have no taste.

Types of Emergence

Probably the most comprehensive description of the types of emergence was created by Fromm [3]. Fromm says that

> Emergence is paradox; emergent properties are often changeless and changing, constant and fluctuating, persistent and shifting, inevitable and unpredictable, dependent and independent from the system in which they arise.

This statement captures the complexity and the difficulty in understanding emergence. The only thing that is not new in this statement is the fact that emergence only occurs in a systems context. Hence, it is a solid systems science concept. Fromm cites the following classifications of emergence:

- The amount of variety in the created system
- The amount of external influence
- Type of constraint maintaining the system's identity (absolute or contingent)
- Number of levels, multilevel of emergence

Taxonomies

Fromm also cites several taxonomies from prior sources. Among these, weak and strong emergence is one of the most common.

- Strong emergence is not deducible from the laws of the lower-level domain. This can be read to mean not reducible to the laws of physics. This taxonomy can be read to mean that strong emergence is generally unpredictable.
- Weak emergence is only unexpected given the laws of the low-level domain. This taxonomy can be read to mean that weak emergence is generally predictable.

As a simple example of weak emergence, powered flight can be seen to be an emergent property of an aircraft. This is true because powered flight cannot be achieved with only part of an aircraft, for example, a wheel.

New Taxonomy

Fromm then proposes a new and more nuanced taxonomy as follows:

- Type I – Simple/Nominal Emergence without top-down feedback
  - Type Ia – Simple intentional emergence, for example, the powered aircraft
  - Type Ib – Simple unintentional emergence
- Type II – Weak emergence including top-down feedback
  - Type IIa – Weak emergence (stable)
  - Type IIb – Weak emergence (unstable)

- Type III – Multiple emergence with many feedbacks
  - Type IIIa – Stripes, spots, bubbling
  - Type IIIb – Tunneling, adaptive emergence.
- Type IV – Strong Emergence

Examples of Emergence

Fromm then provides many examples of emergence in the various categories listed above. These examples may provide useful information regarding the emergent properties the student may see in their investigations. The examples are as follows (by category):

- Weak emergence Category 5.2a (stable)
  - Foraging behavior of ants in colonies
  - Flocking behavior of fish and birds
  - The liquid or fluid property of molecules
  - Self-organization, for example, the World-Wide Web
  - Optimal prices of goods in the economy
- Weak emergence Category 5.2b (unstable)
  - Crashes in the stock market
  - Fads and crazes
  - Social unrest
  - Changes in previously stable norms
  - The bandwagon effect
  - The celebrity effect
  - Buzz in the news
  - Cluster formation in economics
  - Path-dependent solutions (keyboard)
  - Stripes, Spots, Bubbling Category 5.3a
  - Tunneling Category 5.3b
- Strong Emergence Category IV
  - The emergence of life
  - The emergence of culture

HOLISM

Holism has been described in the literature, for example Hitchins [4] that in a complex entity, the whole is greater than the sum of the parts. But what is meant by this statement? We have already discussed the concept of emergence. So, in systems science, the emergent property can be considered an extra *part*, over and above the actual physical parts.

The history of holism can be traced to the history of systems science. The French philosopher Descartes claimed that systems could be considered entities that are the sum of their parts. This idea is called *reductionism* and has largely been rejected by modern science. Reductionism has been replaced by *holism* whose definition is seen above and has been shown in Figure 2.1.

The term was first used by Smuts [5] who maintained that the whole contained all the physical parts and all the activities associated with the whole and the parts. Jackson [6] also added that the whole also consisted of the functions associated with the whole and the parts that are required to perform the functions.

Hitchins [4] points out that holism is an essential principle in Systems Engineering. See Chapter 5 for a comprehensive discussion of Systems Engineering. Hitchins says, "If the system is large, diverse, and complex, it not obvious that changing one part will also change the others, and not necessarily for the better, perhaps only in retrospect." Holism has been described in the literature, for example Hitchins [4] that in a complex entity, the whole is greater than the sum of the parts. But what is meant by this statement? We

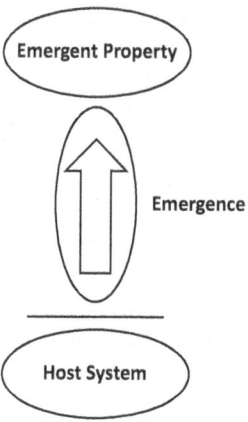

FIGURE 2.2   System Emergence.

have already discussed the concept of emergence. So, in systems science, the emergent property can be considered an extra *part*, over and above the actual physical parts. Figure 2.2 shows how an emergent property can result from a host system.

Hoverstadt [7] goes a step farther. He describes holism to mean "the tendency in nature to form wholes, that are greater than the sum of its parts, through creative evolution." Finally, a question rarely asked in the literature is: what is the value of holism? There is a simple answer that appears to emerge from the writings about it: The value of holism is the ability it provides to see a system in a larger context, for example with its emergent properties included. A second answer to the question of value is that holism helps the analyst deal with systems in which several disciplines are involved, for example, hardware and software.

HIERARCHY

In the bulb there is a flower,
In the seed an apple tree,
In cocoons a hidden promise
Butterflies will soon be free.
[8]

Systems theory was pioneered by Ludwig von Bertalanffy [9] who states that all systems exhibit hierarchy, that is, they can be viewed in layers. He did not mean that they are physically layered. Hierarchy in his view is simply a *view*. Any system in his view, whatever the physical relationship between their parts, can be viewed as a hierarchy. Sometimes these hierarchies may indeed be physical. Other times they may be functional. This concept is described in Chapter 1.

Thus, the use of functional hierarchies is perfectly compatible with the concept of hierarchy. Take, for example, a top-level function. This function can be decomposed into lower-level functions. This step is the beginning of a functional hierarchy, and it is valid regardless of the physical nature of the system.

Common physical hierarchies are beans and other living entities. Inside a bean may be another bean, and so forth. In this case, the hierarchy is indeed physical. But all hierarchies are not necessarily physical. In the case of a bean, when a bean is inside another bean, and so forth, each layer of this hierarchy is called a *holon*. Thus, when a seed contains another seed, and that seed contains a tree, you have the beginning of a holon. Regardless of the origin of the hierarchy of an individual, the hierarchical nature of any system may be depicted as hierarchies.

Documented Hierarchies

A common practice in industry is to depict any system as a hierarchy in documents, such as specifications, and so forth. This practice is not illegal, nor does it violate the definition of a hierarchy. This practice also allows the systems analyst to execute the waterfall model described in Chapter 5. Hence, it can be said that the concept of hierarchy does not invalidate the use of the waterfall model.

In summary, it can be said that the hierarchy is a valid *view* of a system regardless of the physical nature of the system. In addition, various hierarchical products can be generated regardless of the non-hierarchical nature of the physical system.

BOUNDARIES

According to Hoverstadt [7], boundaries "define the identity of the system that you think produces a given set of emergent properties (see Section "Emergence"), so boundaries are key in both modeling and in the creation of emergent properties in reality." If the primary emergent property of an airplane is its ability to fly and you reach the point where it can no longer fly, then you have reached the boundary of that system. [In deference to Professor Hitchins [4], who always insists that an airplane is not a system unless the pilot is included in that system, that definition would apply to emergent properties and boundaries].

Hoverstadt makes another important point: The concept of boundary involves two other laws: The Law of Calling and the Law of Crossing. The Law of Calling involves making a distinction and defining the boundary of interest. The Law of Crossing involves crossing that boundary and going from one state to the other. So, in terms of emergence, the observer is going from a state of emergence to a state of no emergence.

Hoverstadt also discusses boundary setting and emphasizes the difficulty that many boundary setters may have. He does not provide a preferred approach but says that for most cases the answer should be obvious.

REFERENCES

[1] G. Mobus and Michael C. Kalton, *Principles of Systems Science (Systems Science)*. New York: Springer, 2015.

[2] P. Checkland, *Systems Thinking, Systems Practice*. New York: John Wiley & Sons, 1999.

[3] J. Fromm, "Types and Forms of Emergence," Kassel University, Kassel, Germany, 2005.

[4] D. Hitchins, *Systems Engineering: A 21st Century Systems Methodology* (Wiley Series in Systems Engineering and Management). Hoboken, NJ: John Wiley & Sons, 2007.

[5] J. C. Smuts, *Holism and Evolution*. Gouldsboro, ME: Gestalt Journal Press, 1926.

[6] S. Jackson, "Functional Holism of Systems," *System Research and Behavioral Science*, 2022.

[7] P. Hoverstadt, *The Grammar of Systems*. Las Vegas, NV: SCiO Publications, 2022.

[8] N. Sleeth, "In the bulb there is a flower," in *Presbyterian Hymnal*, ed. Louisville, KY, 2013, p. 203.

[9] L. V. Bertalanffy, *General Systems Theory: Foundation, Development, Applications*, Revised ed. New York: George Baziller, 1968.

The Systems Approach

3

Systems Approach is a general term for the process of defining and creating a system using the principles that are generally known as systems theory. The Systems Approach involves two other well-known processes known as Systems Engineering and Systems Architecting. The objective of this chapter is to clarify how these terms have been defined both in official literature and in surrounding publications. Furthermore, this chapter shows how these terms, Systems Engineering and Systems Architecting, can be merged to achieve the Systems Approach.

There are two views of the Systems Approach as it has been used in the past and how it is used today. First, there is the Systems Approach. This term differs from both Systems Engineering and Systems Architecting in that it includes them both (given the narrower perspective on Systems Engineering), and potentially quite a bit more. In other words, it is not possible to define a system without using both systems engineering and systems architecting. Checkland [1] defines the Systems Approach broadly as "a way to tackle a problem" It appears in other sources, including a book by that name by Ramo and St. Clair [2]. According to Maier and Rechtin [3], a true Systems Approach is one that includes the "problem as well as the solution." However, neither one of these sources explains how the Systems Approach relates to Systems Engineering and Systems Architecting. The Systems Engineering Body of Knowledge [4] does that. It concludes that the Systems Approach consists of applying Systems Engineering and Systems Architecting and other processes, such as identifying problems and synthesizing solutions. However, the core processes are Systems Engineering and Systems Architecting as illustrated in Figure 3.1.

The Systems Approach can be seen as the practical modern expression of a utilitarian and rationalist philosophy. It underlies Systems Engineering and much more. It is also foundational for policy formation and is a bedrock of neoclassical/liberal economics. It is built on the belief that things can be understood rationally and quantitatively (if not always rigorously so) that we can divide understanding the truth about situations from judgments about values and "goodness," and that in consequence we can and should divide the processes of understanding problem situations from generating solutions. Finally,

DOI: 10.1201/9781003449935-3

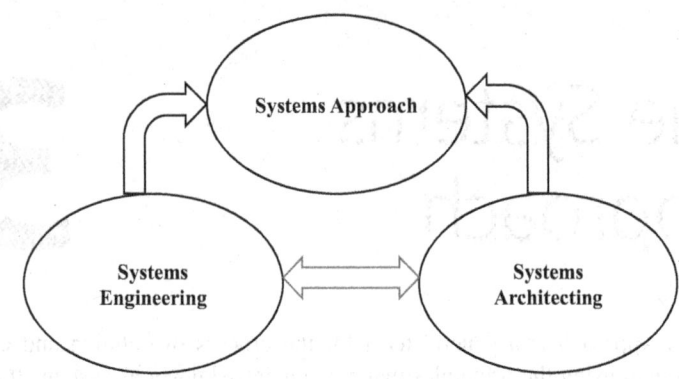

FIGURE 3.1   Systems approach.

having generated solutions, we can and should rationally assess those solutions against problems. The Systems Approach layers onto this philosophical bedrock a recognition that most everything we are interested in is a system and has to be considered that way. We have to look deeply into interactions (as discussed in Chapter 7) and realize that emergent behaviors (as discussed in Chapter 2) are typically the things we care about most, and we have to embrace the inherent complexity of systems.

REFERENCES

[1] P. Checkland, *Systems Thinking, Systems Practice.* New York: John Wiley & Sons, 1999.
[2] S. Ramo and R. St. Clair, *The Systems Approach: Fresh Solutions to Complex Civil Problems Through Combining Science and Practical Common Sense.* KNI Incorporated, 1998.
[3] M. W. Maier and E. Rechtin, *The Art of Systems Architecting*, 3rd ed. Boca Raton, FL: CRC Press, 2009. 1991.
[4] BKCASE Editorial Board. "Systems Engineering Body of Knowledge (SEBoK)." http://sebokwiki.org/wiki/Guide_to_the_Systems_Engineering_Body_of_Knowledge_(SEBoK) (accessed 2016).

Systems Engineering

4

THE SYSTEMS ENGINEERING VEE MODEL

Figure 4.1 shows the well-known systems engineering Vee model. The Vee Model captures all the important aspects of systems engineering. For example,

- It shows that the requirements process starts at the top of the system hierarchy with stakeholder needs. See discussion above for a description of stakeholder needs. Remember the concept of hierarchy from Chapter 2.
- It shows that requirements are flowed down through the system hierarchy. See Chapter 2 for a discussion of the concept of system hierarchy. The Waterfall model represents the flow-down of requirements through the system hierarchy.
- It shows that the system reaches a state of synthesis when all requirements are derived to the lowest level of the system hierarchy. Also, see Chapter 2 for a discussion of the concept of hierarchy. See the Systems Engineering Management section, below, for a discussion of synthesis.
- It shows that the entire system reaches a state of verification when all elements have been verified to the top level of the system hierarchy.
- It shows the progress of system development through all levels of requirements and verification. See once again the concept of hierarchy as before.

Functional Analysis

This is a vital part of the systems engineering process. Functionality is an essential element in establishing the purpose of the human-made system. Functions have an important role in systems theory and in systems application.

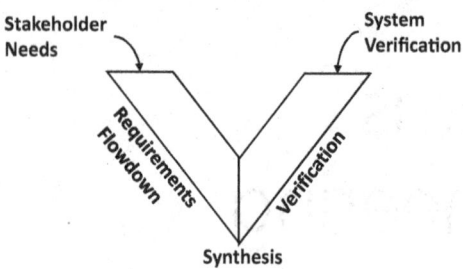

FIGURE 4.1   Vee model.

In systems engineering, functions have the role of creating functional requirements for the elements of the system. That is, they will assist in determining how well an element must perform its function. Also, as will be seen in
the chapter on Systems Architecting through Functions (Chapter 5), functions
will be seen as an important aspect of creating the architecture of the system.
Functions also play an important role in defining the subsystems and how they
relate to each other.

Requirements and Needs

Stakeholder needs are the beginning of the systems engineering process.
However, this is the area where flaws can occur especially when stakeholders
are not familiar with Systems Engineering. It is the role of systems engineers to
determine whether stakeholder needs are logical and achievable. If not, mana-
gerial steps should be taken to resolve the issues.

   a. Requirements are both derived from Stakeholder Needs and from
      Functional Analysis·(see above).
   b. Requirements are also flowed down using the Waterfall Model as
      described in the Hierarchy section of Chapter 2.
   c. Requirements are verified by test and other methods as described
      below.

ALL-UP TESTING

It is generally agreed that all-up testing can take responsibility for the
success of the Apollo program. According to [1], this type of testing was
introduced to NASA by George Mueller. This type of testing involves
the testing of an entire system including its subsystems in its operational

configuration, for example, a NASA space vehicle and not simply individual components. Its purpose is to validate the functionality and performance of an entire complex system, subsystem, or product.

This type of testing shows not only that all components worked, but rather that they all worked *all together* seamlessly, including both hardware and software systems and subsystems. Examples of domains that may undergo all-up testing include aircraft, spacecraft, automobiles, and large-scale software applications. It is assumed that due to the success of this type of testing, other industries are also adopting it.

In addition to its other ambitious testing goals, this type of testing helped NASA meet its end of the decade goal for landing on the moon and returning safely to Earth. NASA had originally estimated that it would take more than 20 years to accomplish the moon landing. While complete system testing can be more time-consuming and resource-intensive than testing individual components separately, it provides a more accurate assessment of how the system and its subsystems will perform under realistic operational conditions and can help to identify potential problems early in the development process, when they are easier and less expensive to repair. In addition to full system testing, NASA performed many partial system tests for risk reduction.

Constraints and Specialty Requirements

The development of a system normally requires the implementation of many constraints and specialty requirements. Reliability is an example. When reliability requirements are established, their execution is handled in the same way as other requirements. That is, they will appear in the procurement specifications for the various components of the system.

Interfaces

Interfaces are a key factor in the holism of any system. Human-made systems consist of many parts, and many of the parts interact with each other. This fact gives rise to the process known as Interface Management. This process controls how the parts interact with each other. Interface Management includes the management of the requirements between parts and also the physical interface between the parts.

Figure 4.2 shows how both external and internal interfaces are treated in systems engineering. Both external and internal interfaces have both requirements aspects and physical aspects.

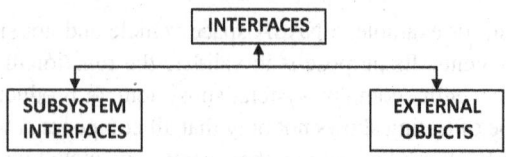

FIGURE 4.2   Interfaces.

Verification and Validation

The difference between verification and validation is not clear to many practitioners.

   a. Requirements validation is the proof that a requirement is valid.
   b. Requirements verification is the proof that a design meets a requirement. There are four methods: test, demonstration, analysis, and inspection. See section "Requirements and needs" on All-up testing for an explanation of how important tests may be.
   c. System validation is the proof that a design meets customer needs. This step normally requires the participation of the customer.

Design reviews should cover all aspects of verification and the success or failure of tests and analyses. The previously mentioned all-up testing satisfied the verification requirement.

Validation of the System

   d. Design reviews should cover the status of the validation of the system, that is, does the system meet the needs of all stakeholders?

Systems Engineering Management Plan (SEMP)

The SEMP is an essential tool for assuring that a project has been executed as planned.

First, the SEMP must be approved by the Project Manager and followed and tracked by him or her. Typical items in the SEMP are as follows:

   • Requirements management plan
   • Planned major reviews, audits, and other control points
   • Risk management plan
   • Safety and certification plan

- Verification plan
- Decision management plan
- System resilience plan

System Resilience

System Resilience has become a part of systems engineering partially influenced by the book *Resilience Engineering: Concepts and Precepts* [2]. The INCOSE Resilient Systems Working Group has defined system resilience as "the ability to maintain capability while facing adversity." A common misunderstanding of system resilience is that it does not mean the ability to return to a prior state but rather to maintain a desired capability.

System Resilience Techniques

Following the founding of the INCOSE working group in 2007, the group immediately began to work on identifying techniques to implement the resilience of a system. For clarity, only human-made systems were considered.

The identified techniques were organized into a framework called taxonomy. The taxonomy is just one of several Knowledge Organization Systems (KOSs). The taxonomy is one of the most common KOSs.

The resulting taxonomy is shown in the SEBoK. Layer 1 of the taxonomy reflects the intrinsic value of resilience. There are three items in Layer 1: (1) avoidance, (2) withstanding, and (3) recovery. Layer 2 reflects the means objectives which are not ends in themselves. There are 18 items at this level. Layer 3 is the collection of architectural techniques for achieving resilience. There are 38 items at this level. Members of the group have published analyses for the derivation of these techniques. For example, at least one PhD thesis [2] has been written on the subject. Another paper by Brtis [3] published one paper

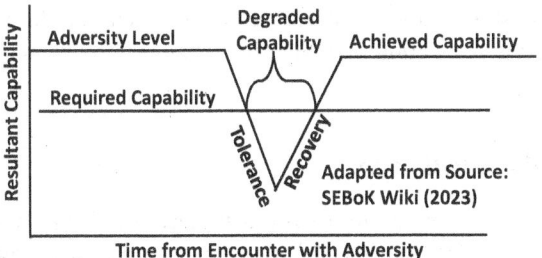

FIGURE 4.3   Resilience timeline.

applicable to resilience in a military context. The results of these publications are found in Layer 3 of the Taxonomy.

The System Resilience Process

Following is a summary of the system resilience process:

- Mission analysis: Identify problem space including adversities and solution classes.
- Stakeholder needs: Identify resilience scenarios.
- System requirements: Identify resilience-related requirements.
- Architecture definition: Identify candidate architectures.
- Design definition. Identify candidate designs.
- Risk identification. Identify risks associated with resilience choices.

Resilience Requirements

Following is a list of requirements for resilience:

- Operational concept name
- System of interest
- Capability of interest (see Figure 4.2)
- Required amount of capability
- Modes of operation, for example, training
- System states during operation (see section on System States)
- Adversity type
- Stresses on system
- Scenario constraints, for example, cost
- Resilience metrics (see Eq. 1)
- Natural language pattern of resilience, for example, mode, state, adversity, stress, capability, scenario, and metric.

Figure 4.3 illustrates how the capability of a system can be maintained or improved by the imposition of resilience techniques following the encounter with an adversity.

Affordable Resilience

Affordable resilience means to achieve a balance across Life Cycle cost and Technical attributes. These include providing capability across changing conditions. Technical attributes include addressing risks and achieving required techniques.

Discipline relationships

Resilience has many aspects in common with other quality processes including availability, environmental impact, survivability, maintainability, reliability, operational risk management, safety, security, and quality.

Human Considerations

Human actions are an important part of many resilience techniques. Humans make decisions at their own discretion. Apollo 11 is a good example.

Metrics

Brtis has found that the following metric is useful across many systems and scenarios.

$$R = \sum_1^n \left(\frac{P_i}{T} \int_0^T Cr(t)_i \, dt \right)$$

where

R = Resilience of the required capability (Cr);
n = the number of exhaustive and mutually exclusive adversity scenarios within a context (n can equal 1);
P_i = the probability of adversity scenario i;
$Cr(t)_i$ = time wise availability of the required capability during scenario i: 0 if below the required level, 1 if at or above the required value. Where circumstances dictate this may take on a more complex, non-binary function of time;
T = length of the time of interest.

Resilience Heuristics

A set of heuristics have been found to be useful in designing a resilient system. These heuristics can be seen to be part of Architecting through Heuristics. They are, therefore, found in Chapter 6.

Risk Analysis and Management

Attention to risk is one of the most difficult to achieve in industry. A common culprit is *risk denial* which is the refusal to admit that a project has any risks at all. In truth, most projects have many risks, and if the risks are not recognized, they will not be mitigated. A notable authority on risk is [4]. Vaughn [5] is also recognized for her analysis of the *Challenger* event. The classic definition of risk is the product of probability [of the risk occurring] and the consequence of its occurrence. The estimation of probability is another challenge. Risk can be mitigated (by test or other method), accepted (if the probability or consequence) is very low, or the risk is transferred (to another agency).

However, considering the number of high-consequence failures (*Columbia, Challenger,* and others), the need for risk management should be obvious. Probably one of the most authoritative sources of information on risk mitigation is the Columbia Accident Investigation Report [6]. This report recommends that no major action should be taken without approval from an independent technical authority (ITA). The key word is *independent*. This word means that the authority is both organizationally and financially independent of the agency performing the risky action.

Design Reviews

Design reviews are a common method of assuring that a project is progressing as planned. A common mistake is to hold design review meetings even when the goals for that meeting have not been met, for example, requirements development and design progress. It is important to hold these meetings only after goals have been met.

The System Life Cycle

Human-made systems are normally intended to last a long time. For that reason, the concept of System Life cycle exists. The purpose of the System Life Cycle is to analyze all phases of the Design reviews should cover all phases of the system life cycle including its conception, its development, its testing, its operation, and its disposal.

Trade Studies

Design reviews should cover the status of design trade studies, that is, do subsystems meet their intended goals?

Synthesis

Design reviews should cover the design status of the entire system, that is, does it meet the intent of the stakeholders? Synthesis occurs when requirements are brought together through trade studies. This process occurs at the end of the system hierarchy. The synthesis process determines how the configuration is formed at the end of the requirements process.

Decision Management

Decisions with biases. Many texts assume that all decision makers can think logically if they want. They just have to will it. The Systems Engineering Handbook [7] shows that most decisions are made with biases, that is, decisions with presumed outcomes. These biases should be accounted for in all decisions, whether they be design decisions or operational decisions.

Certification and Safety

The generally acknowledged authority on Safety is Leveson [8, 9]. Later in this book (Chapter 8) we examine how artificial intelligence (AI) might have the potential for performing human functions centered on enhancing the safety of any system, in particular those systems who depend on human actions to perform critical functions.

A system is certified as safe when it meets specific requirements to show that it is free from failure under specific conditions. A recognized authority on safety is [8, 9]. The safety process is a subset of the entire Systems Engineering process.

Systems Engineering Heuristics

This section presents a selected group of heuristics that have been collected within the International Council on Systems Engineering to help practitioners develop new and effective systems. The heuristics can be found at the website www.INCOSE.Org/heuristics. Heuristics are short-hand statements of tasks that enable the systems engineer to design a system.

The first three heuristics pertain to practices performed by systems engineers throughout the life cycle of the system. These practices include normal systems engineering activities, such as requirements development and flow-down. These practices will be reflected in the Vee model previously described in this chapter. Developing the system in stages is a normal systems engineering practice to minimize the risk of design errors during development. The use of subject matter experts during these phases is always a good policy. The use of realistic scenarios to validate the design is also a good practice. These scenarios can be used either in live testing or in simulations. What will be found in these heuristics is a call for caution in their execution. The following two references are excellent sources of heuristics, [10, 11].

TABLE 4.1   Systems engineering heuristics

TOPIC	CHAPTER	HEURISTIC
Systems engineering practices	3	Developing the system design in stages can reduce risk if the critical system behaviors are designed first.
		Seek external subject matter experts when in-house knowledge is lacking.
		Use cases or scenarios illustrate intended operation of the system, so they must be realistic to be effective.
		Only insist on requirements that are essential.
		Thoroughly understand the problem or issue the customer needs to solve, not just surface-level symptoms.
Systems engineering artifacts	3	Tailor your systems engineering information artifacts around the decisions they are intended to support, and the people who will be making the decisions.
Systems engineering focus	3	Be part of the solution, not part of the problem.
System functions and attributes	3	Function (what the system does and to whom/what) as well as key attributes (how well it does it) need to be explicit in and understood from the model.
System boundaries	2	Choosing system boundaries carefully is critical to systems engineering success.
Systems engineering terms	3	Establish a lexicon of key systems engineering subject matter terms.
System perspectives	3	Optimization of any single function constrains accommodating all the different perspectives.

This heuristic emphasizes the importance of verification, especially testing, and warns against the use of expensive testing when it is not needed.

Verification	3	Testing is expensive—make it efficient

Chapter 3 emphasizes the importance of interfaces. This heuristic call for an owner of each interface.

Interfaces	3	Each interface requires an owner.

The following heuristic reflects the wide use of tools on systems engineering projects and the importance of the user of these tools to be competent in their use.

Use of tools	3	Tool users should be competent in the principles and processes that are supported by the tool.

The following heuristic reflects the wide use of ilities (reliability, etc.) and the importance in the execution of those ilities.

Ilities	3	For suitability, such as safety and each other relevant "ility," it is important to articulate what the minimal acceptable level is, as well as the level beyond which little extra value is added or is feasible.

The following heuristic reflects the importance of good communication among project team members and with the customer.

Communication	3	Understand that when you communicate with people, they will interpret what you communicate using their own mental models.

The following heuristic reflects the importance of knowing failure modes and taking them into account in system design.

Failure modes	3	Design the "weak links" in the system to fail on purpose at appropriate times.

The following heuristic reflects the fact that component reuse is common in industry. This heuristic cautions the designer against reusing components when such an action may result in a loss of system performance.

Component reuse	3	Reuse can impose constraints on system performance.

The following heuristic cautions the designer against making design changes without a justification for doing so.

System changes		Don't accept assurances that there will be no changes.

The following heuristic cautions the designer to account for the culture of the project at hand. For example, commercial and military cultures may be very different.

Culture	3	A key systems engineering success factor is ensuring that the project culture is aligned and compatible with the development strategy.

The following heuristic advises the designer to maintain a configuration management process throughout the development process and to make the system model compatible with the managed configuration.

Configuration management	3	Configuration management of the system model is configuration s paramount.
Simple problems	3	As we learn, do not be overly attracted to potentially simple problems or solutions without due diligence.

The management of complexity is a high-priority task in system development. The following heuristics provide advice on how to do that. Accounting for culture in these tasks is important. Consideration of unexpected emergent properties is also important. The large number of complexity heuristics reflects the importance of this topic and the difficulty in dealing with it.

Complexity	3	Think big, assess all, but start small and continuously evolve when handling complexity.
		Organizational culture and structure must adapt and adjust to deal with complex systems or environments.
		Engineers can aggregate and iterate more rapidly with stable intermediate forms (or increments) of the design.
		During the development and evolution of complex systems, create options and keep them open for as long as possible.
		Complex systems should be made as simple as possible to deliver all needed system outputs, but not simpler.
		An element good enough in a previous system needs careful analysis before being applied to another complex system.
		Proactively and holistically manage the unexpected emergent issues within a complex system to mitigate failure.
		It is important to understand, recognize, and characterize complexity.

For complex problems, focus on holistic utility, not optimization of any single objective.

Strategically assess the areas of complexity to select handling approaches thoughtfully.

Proactively and holistically manage unexpected emergence within complex systems to mitigate failure and realize opportunity.

Apply iterative approaches to developing or evolving a complex system.

It is counterproductive to merely label a system as complex or not complex.

Understanding your organization as a complex adaptive system is critical to incorporating improved systems engineering.

Observe a system from multiple perspectives to determine complex characteristics.

Do not assume successful techniques for complicated systems will be helpful with a complex system.

Complexity necessitates continuous learning and adaptation.

Further Reading

This chapter has managed to describe some of more important aspects of Systems Engineering, and in addition, it discusses some of the more recent developments in this field including:

- System Resilience
- Systems Engineering Heuristics, and
- All-up testing. However, if the reader needs exhaustive discussion of Systems Engineering, we recommend the following sources: Sage and Armstrong [11] or the INCOSE Handbook [7].

We hope you will find these sources satisfactory.

REFERENCES

[1] NASA. "All-Up Testing." https://appel.nasa.gov/2010/02/25/ao_1-7_f_snapshot-html/ (accessed 25 August 2023).

[2] E. Hollnagel, D. D. Woods, and N. Leveson, *Resilience Engineering: Concepts and Precepts*. Aldershot, UK: Ashgate Publishing Limited, 2006, p. 23.

[3] J. Brtis, "How to Think About Resilience in a DoD Context," MITRE Corporation, Colorado Springs, CO, August 2016.

[4] E. H. Conrow, *Effective Risk Management: Some Keys to Success*, 2nd ed. Reston, VA: American Institute of Aeronautics and Astronautics, 2003.

[5] D. Vaughn, *The Challenger Launch Decision: Risky Technology, Culture, and Deviance at NASA*. Chicago, IL: University of Chicago Press, 1997. 1996.

[6] NASA, "Columbia Accident Investigation Report," National Aeronautics and Space Administration (NASA), Washington, DC, August 2003, vol. *1*.

[7] INCOSE, "Systems Engineering Handbook," International Council on Systems Engineering, Seattle, CA, Handbook January 2015.

[8] N. Leveson, *Safeware: System Safety and Computers*. Reading, MA: Addison Wesley, 1995, p. 434.

[9] N. Leveson, *Engineering a Safer World; Systems Thinking Applied to Safety*. Cambridge, MA: MIT Press, 2012.

[10] M. W. Maier and E. Rechtin, *The Art of Systems Architecting*, 3rd ed. Boca Raton, FL: CRC Press, 2009. 1991.

[11] A. Sage and J. E. Armstrong, *Introduction to Systems Engineering*. Hoboken: Wiley, 2000.

Architecting through Functions $\mathbf{5}$

This chapter illustrates concepts, examples, and exercises in the architecting process considering the benefits and importance of systems theory applications "all the serious mistakes are made in the first day," according to Maier and Rechtin [1].

"Generally speaking, engineering deals almost entirely with measurables using analytic tools derived from mathematics and the hard sciences." According to Rechtin [2], this concept has come to be known as Systems Engineering.

"Architecting deals largely with un-measurables using non-qualitative tools and guidelines based on practical lessons learned." According to Rechtin [2], this concept has come to be known as Systems Architecting. Traditionally, Systems Architecting is performed before Systems Engineering on the assumption that the arrangement of the parts of the system should be determined by how well all the parts fit together. In practice, there is a lot of overlap between the two disciplines. This chapter will show that both disciplines depend on functions.

All the architecting processes normally need to ask basic questions: "What solution do we wish were available? What is difficult with the current product we use?; and Why does it not do something we want it to?" These are the basic questions posed by Maier and Rechtin [1, p. 17].

The answers to these forming a vision help to analyze and understand what the stakeholder wants the new product to do. According to Blanchard, "defining the problem is the most difficult part of the SE process" [3].

Systems Architecting is a branch of the Systems Approach that has come to the fore in recent years. The reason that this field is seen as separate from Systems Engineering is that it is seen as more qualitative than systems engineering. However, both Systems Engineering and Systems Architecting are required to define a system, as stated by Jackson and Moraes [4].

DOI: 10.1201/9781003449935-5

Systems Engineering has a focus on performance, while Systems Architecting has a focus on the architecture (both physical and functional) of the system. Rechtin [2] was a pioneer in the development of Systems Architecting. Other resources are Maier and Rechtin [1], Sillito [5], and Jackson and Moraes [4].

According to BKCASE Editorial Board [6], both Systems Engineering and Systems Architecting are two of the processes required to define a system through the Systems Approach among other processes as stated by Jackson and Moraes [4].

According to Rechtin [7], the value added by the system as a whole, beyond that contributed independently by the parts, is primarily created by the relationship among the parts; that is, how they are interconnected. We saw in Chapter 1 that a primary feature of a system is the interactions among its parts.

DETERMINING THE ARCHITECTURE

There are two primary ways of determining the architecture of a system, the heuristics method, and the functional analysis method. In practice both methods may be used for a single system. In this book, the functional analysis method is shown as an example.

THE HEURISTICS METHOD

As explained by Rechtin, heuristics are broadly stated objectives that outline what is to be done. The first is the method outlined by Rechtin [2]. It is called the heuristics method. In short, the heuristics method establishes the parts of the system and the arrangement of these parts through the application of heuristics.

THE FUNCTIONAL ANALYSIS METHOD

Functional analysis is the systematic process of identifying, describing, and relating the functions a system must perform in order to be successful. It does not address how these functions will be performed according to NASA [8].

In the early phases of the project life cycle, functional analysis deals with

- The top-level functions that need to be performed by the system;
- Where these functions need to be performed;
- How often they need to be performed; and
- Under what operational concept and environmental conditions.

Within the functional analysis process, we start with understanding customer needs and operational needs extracted from the concept of operations (CONOPS). This first step, called a functional survey, where we managed to map the first correlations of needs, transforming them into possible actions through a grammatical structure using verbs in the infinitive, thus declaring the functions for a proposed solution. Table 5.1 illustrates how functions can be allocated to different elements of the system. This table shows that the architecture of a system is driven by the functions of its parts.

Here we can bring an example where we have both operational customer needs and where we can derive functions based on these needs. This process is illustrated in Figure 5.1.

For this example, we use an aircraft entertainment connectivity need as mentioned below, but the same structure to deploy this need in this example can be performed to other needs that we mapped on the Stakeholder Needs and Requirements definition process in the technical Systems Engineering process.

User requirements definition phase, in the first phase the role of architect is to reduce ambiguity and to elicit needs and try to keep the architecture as solution neutral as possible. This phase is often called the "User Needs Phase." This name emphasizes the fact that the requirements identified in this phase are not design requirements prioritizing high-level requirements that result from the goals; the requirement is more related to what my system will do to meet the user's needs. The needs of stakeholders and beneficiaries guide the entire process; every decision should be traced back to them. The requirements

TABLE 5.1 Typical allocation of functions to system elements

FUNCTION	SYSTEM ELEMENT
Lift	Wing
Thrust	Engine
Connectivity	Communication System
Roll	Ailerons
Lateral Control	Vertical Control Surfaces
Pitch Control	Horizontal Control Surfaces

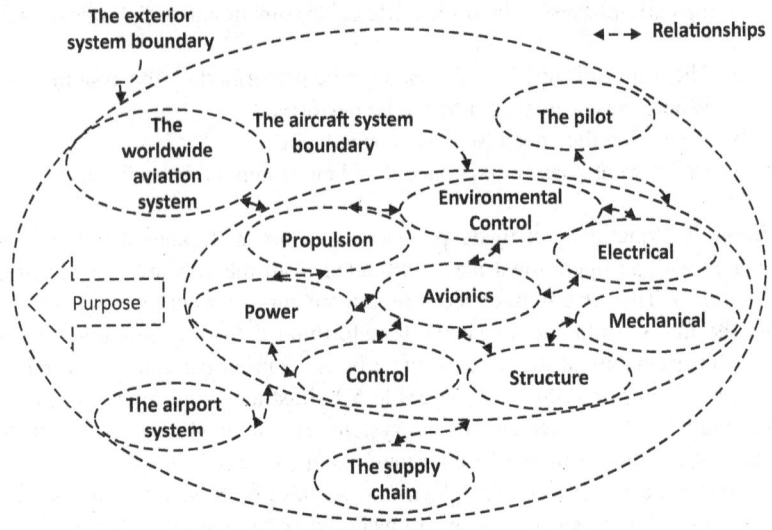

FIGURE 5.1   Sub-System Architecting.

identified in this phase are tradeable throughout the development process. So that is more important to the user as stated by Blanchard [3].

Let's look at some questions based on the proposed need case below and the steps to help us perform our running example.

- Example Need: Connectivity.

  How can we understand what really needs to be in the solution so that we can satisfy it? Always start with the question why? Why do you need connectivity?

  To answer the question below, we will use context analysis (Evaluation of the operational scenario—Mission Analysis Process) to have a better understanding of this need, thus allowing us to write an understanding for our example.

- In this case, one possible scenario (context) to this need to user is: "Possibility to be productive during the flight with connectivity."

  As a systems architect, it is the moment that we define what our proposal can have and perform (functions) that satisfy this need of our exercise. Here are some proposals using once again the questioning method: What should my solution provide to meet the need?

- What should the solution provide to meet the need?
  - The associated function is: Access to the network of computers dispersed throughout the planet that exchange data and messages using a common protocol (Internet).

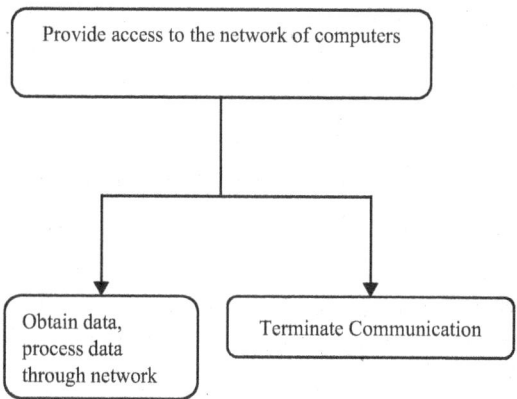

FIGURE 5.2 Functional decomposition.

Responding to the question above, that is, we define the function at the system level where we are talking about what and not how that proposed solution will be implemented. This is an important difference, so we don't get into implementations in the early stages of the life cycle as described in [9].

Functions are related to the capabilities of delivering perceived value to the customer and will be deployed in another stage of our process that we call functional allocations. Once this is done, then we have the need already with a context and a capacity (function) offered that satisfies the need. From this capability, we can already have our first high-level requirements, called HLRs, like this in our example:

- HLRs (High-Level Requirements): The system shall be capable of providing access to a network of computers spread across the entire planet that exchange data and messages (Internet) during the flight experience.
- Note, that it is still an HLR, that is, the verification level is related to providing or not providing yet, without performances associated with the requirement, thus seeking at this point in the life cycle (Conception), the perception in a validation of these HLRs by the client, if the path proposed by the architect to the client is going the right way.

We have arrived here at the end of a first level of functional structure in our journey, where we have a clearer understanding of the meaning of this need and a line of thought of what we can offer my client to satisfy his need, in addition to having already proposed HLRs derived from need.

In the next topic, we deploy this system-level function to subsystem-level functions using questions method and Functional Flow Block Diagrams (FFBD).

Later in these early phases, functional analysis proceeds to lower levels of the system decomposition to define the system's functional design and interfaces.

- To draw out all the functions the system must perform to meet its requirements.
- Required for subsequent requirements allocation.
- To identify profitable trade studies.
- In describing what must be done and not how to do it, we decouple requirements from implementation. This leaves the implementation trade spaces unbiased.

But "…decomposition cannot be found in a book or dictated by the stakeholders; the decomposition is a product of the engineers of the system and is part of the architecture design process that is attempting to solve the design problem established by the requirements. The decomposition can be carried out as deeply as needed to define the transformations that the system must be able to perform, as described by Buede [10]."

Let's look at a running function example and decompose it: For the second level of functions, we use the question method and ask: How to break into action steps to: "access to the network of computers dispersed throughout the planet that exchange data and messages using a common protocol (Internet)," let's look at a possibility below. The flow-down of these functions is shown in Figure 5.2.

- Obtain data,
- Process data,
- Authenticate credential,
- Validate data,
- Distribute data,
- Terminate communication.

Then, Figure 5.3 represents our example function decomposition. A primary functional analysis technique is the FFBD, the idea is to show the sequential relationship of all functions that must be accomplished by a system.

FFBDs have the classic features of logic structures and control flow. Part of the behavioral (logical architecture) representation set, they unambiguously represent the flow of control through sequencing of functions and constructs, focusing exclusively on structural control and ignoring the sequencing and synchronization aspects of the corresponding data flow.

Each function (represented by a block) is identified and described in terms of inputs, outputs, and interfaces from top down so that sub-functions are recognized as part of larger functional areas.

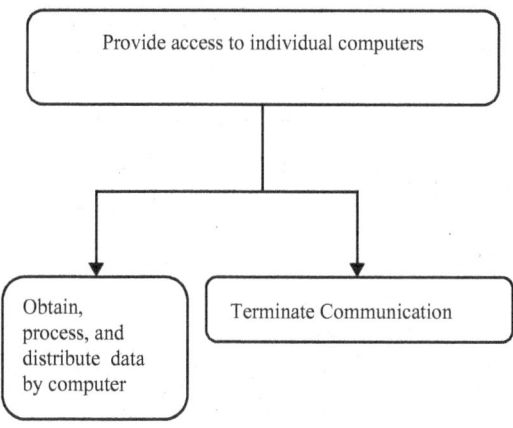

FIGURE 5.3 Second-level decomposition.

The FFBD network shows the logical sequence of "what" must happen and does not assume a particular answer to "how" a function will be performed as described by Buede [10].

In order to allow the sequence of functional decomposition, we will use the FFBD method for the first-level decomposition in Figure 5.3 to demonstrate the second-level decomposition.

This figure shows how the system (the aircraft) communicates with both interior and exterior nodes. Some functions may be performed in parallel, or alternate paths may be taken. The functional analysis is then expanded as shown in Figure 5.4.

Functional analysis is implementation independent. All functions are described in terms of what must be done (and sometimes how well) not how it will be done. This independence ensures that when subsequent trade studies choose how functions will be performed they will be unbiased. Summary: Functional Analysis in the Architecting process: Buede [10].

- Functional analysis is a system development tool used to capture required system functions.
- Functional analysis also supports functional decomposition—the process of describing the sub-functions that are necessary for each function.
  This figure shows that subsystems need to be architected within the system as a whole.
- FFBDs are graphical tools used to capture the functional sequence and functional hierarchy of a system.

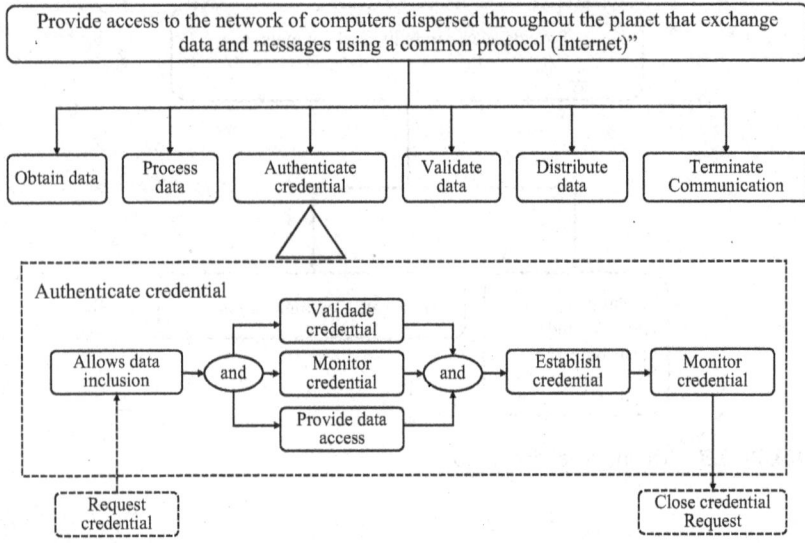

FIGURE 5.4   Credential authentication.

ALTERNATIVE AIRCRAFT ARCHITECTURES

The standard architecture for commercial aircraft has remained unchanged for many decades. It is called the two-surface (wing plus stabilizer) architecture. Early aircraft including the Wright Flyer had a canard surface in a forward position, but that concept was discarded in favor of a rear stabilizer because the latter provided improved stability.

More recently, other alternative architectures have been explored with the hope that they will provide increased performance. They are the blended wing-body and the three-surface architecture. These concepts will be discussed in the following sections. In addition, another concept will be described that provides superior landing and takeoff performance.

BLENDED WING-BODY (BWB) ARCHITECTURE

The blended wing-body [11] is an architecture that promises to provide improved performance. The main architectural feature of the BWB is that it has no fuselage, or it may be said that the fuselage is integrated into the wing. Both passengers and cargo are held within this wing-fuselage. The argument for this design is that by eliminating the fuselage, a considerable reduction in aerodynamic weight will be realized. In addition, since the whole aircraft is a wing, a significant improvement in lift will be realized. Analysis shows that this architecture will result in a considerable increase in cargo capability and in overall aircraft range compared to a conventional aircraft [12].

Overall Pros and Cons

Our newly established database allowed us to identify the main obstacles of the BWB architecture, mostly related to airfoil selection and evolution along the aircraft, chosen materials, structural concerns, stability issues, lift-to-drag ratio optimization, manufacturing, and transonic phenomena [13].

System architecting is the process of developing different architecture configurations for a given aircraft system. This is done through an activity called Design Space Exploration (DSE) which constitutes the enumeration, representation, and evaluation of all possible architectural design solutions, for example BWB. This process identifies all possible combinations of system components to develop a list of candidate system architectures. Feasible system architectures are then evaluated, and a solution architecture that satisfies the requirements is found [14]. Systems architectures need to be comprehensively explored during conceptual design for several reasons:

1. To ensure that the chosen systems architecture satisfies requirements and is compatible with the aircraft configuration.
2. To ensure that any integration and installation effects are considered early in the design process to avoid costly rework and redesign.
3. To ensure that safety considerations are addressed early in the design process.

Architecture evaluation comprises activities aimed at eliciting system architecture performance features such as safety, cost, and weight. For instance, the power requirements for a system are driven by system sizing, which is in turn influenced by the aircraft configuration and its mission. Architecture evaluation therefore determines system size, power requirements, and the effect of the system on overall aircraft performance.

Functional analysis prescribes a hierarchical approach of function decomposition from system to subsystem-level functions. This allows a link between the top-level requirements to the system architecture at various levels of abstraction. Functional analysis is therefore a fundamental tool in DSE and architecture definition and very useful on the BWB architecture decisions [15].

THREE-SURFACE AIRCRAFT

Another architectural feature that promises to improve the performance of commercial aircraft is the three-surface (wing-stabilizer-canard) architecture [16]. Although this concept has not received much attention in recent years, analytically it does promise to improve performance. The rationale behind its design is as follows.

Conventional aircraft normally have two surfaces, the wing and a stabilizer. This architecture does not allow the aircraft to fly at both a maximum lift (maximum payload) and optimum angle of attack (maximum range) simultaneously. A third surface, a canard, will allow the aircraft to do that. It is normally conceived that this third surface would be located in the vicinity of the cockpit.

There are some drawbacks to this architecture: First, an additional surface will add both weight and control systems to the overall aircraft. It all depends on whether the improved performance exceeds the losses due to extra weight. Hence, a trade study will be required to do this.

MILITARY CARGO AIRCRAFT

There is a cargo aircraft in the US inventory that does have some architectural features that offer distinct benefits. Although these features are unlikely to be incorporated into commercial aircraft design, it is worth mentioning them here to demonstrate the benefits of architectural features. This aircraft is called the C-17. The first architectural feature to be incorporated on this aircraft is a high-lift thrusting configuration. With this feature the wing flaps are subjected to exhaust from the engines and vectored down to produce increased lift during takeoff and landing. This feature has resulted in significantly shorter takeoff and landing distances. To accomplish this, the flaps require improved insulation to withstand the exhaust from the engines.

The second architectural feature is the capability to carry a significantly higher amount of fuel to achieve increased range. After a significant amount of analysis, it was decided to add a new fuel tank inside the fuselage since there was no room in the wings, the traditional place to store fuel. So, this new fuel tank in the fuselage was the architectural change that achieved the range increase and also has a secondary barrier in the added fuel tank to prevent fuel from leaking on the cargo.

REFERENCES

[1] M. W. Maier and E. Rechtin, *The Art of Systems Architecting*, 3rd ed. Boca Raton, FL: CRC Press, 2009. 1991.

[2] E. Rechtin, *Systems Architecting: Creating and Building Complex Systems.* Englewood Cliffs, NJ: CRC Press, 1991.

[3] B. Blanchard and W. J. Fabrycky, *Systems Engineering and Analysis*, 4th ed. (Prentice Hall International Series in Industrial and Systems Engineering). Upper Saddle River, NJ: Prentice Hall, 2006. 1981.

[4] S. Jackson and R. Moreos de los Santos, *Systems Approach to the Design of Commercial Aircraft.* Boca Raton, FL: CRC Press, 2021.

[5] H. G. Sillitto, "System resilience fundamentals," presented at the *INCOSE IS 2016*, Edinburgh, Scotland, 2016, Presentation.

[6] INCOSE, "Systems Engineering Handbook," International Council on Systems Engineering, Seattle, CA, January 2015.

[7] E. Rechtin, *The Systems Architecting of Organizations: Why Eagles Can't Swim.* CRC Press, 1999.

[8] NASA, "Columbia Accident Investigation Report," National Aeronautics and Space Administration (NASA), Washington, DC, August 2003 2003, vol. *1*.

[9] BKCASE Editorial Board. "Systems Engineering Body of Knowledge (SEBoK)." http://sebokwiki.org/wiki/Guide_to_the_Systems_Engineering_Body_of_Knowledge_(SEBoK) (accessed 2016).

[10] D. M. Buede, *The Engineering Design of Systems (Wiley Series in Systems Engineering).* Hoboken, NJ: John Wiley & Sons, Inc., 2000.

[11] A. Bowers, "Blended-Wing-Body: Challenges for the 21st Century." NASA Dryden Flight Research Center. http://www.twitt.org/BWBBowers.html (accessed January 9, 2014).

[12] O. V. Salazar, *Development of blended wing body aircraft design.* Département de génie mécanique, École de Technologie Supérieure, Quebec, 2015.

[13] N. Viola, S. Corpino, M. Fioriti, and F. Stesina, "Functional Analysis in Systems Engineering: Methodology and Applications," in *Systems Engineering – Practice and Theory*, pp. 71–96.
[14] Andrew Kingsley Jeyaraj, *A Model-Based Systems Engineering Approach for Efficient System Architecture Representation in Conceptual Design: A Case Study for Flight Control Systems*. Concordia University Montreal, Quebec, Canada, 2019.
[15] E. L. Cole, "Functional Analysis: A System Conceptual Design Tool [and Application to ATC System]," *IEEE Transactions on Aerospace and Electronic Systems*, vol. *34*, no. 2, pp. 354–365, 1998.
[16] AIAA, "Three-surface Aircraft," 2023. [Online]. Available: https://arc.aiaa.org/doi/10.2514/3.45827

Architecting through Heuristics

6

THE HEURISTICS METHOD

As explained by Rechtin [1], heuristics are broadly stated objectives that outline what is to be done. The first is the method outlined by Rechtin. It is called the heuristics method. In short, the heuristics method establishes the parts of the system and the arrangement of these parts through the application of heuristics. What is a heuristic? Some examples of heuristics are:

- Rules of thumb
- Best practices
- Common sense
- Intuitive judgments

This chapter will bring some examples of the use of heuristics to aid the Architecting process based on Maier and Rechtin [2].

Examples of heuristics are:

1. Don't assume that the original statement of the problem is necessarily the best or even the right one.
2. In partitioning, choose the elements so that they are as independent as possible; that is, elements of low external complexity and high internal complexity.
3. Simplify. Simplify. Simplify.
4. Build in and maintain options as long as possible in the design and implementation of complex systems. You will need them.
5. In introducing technological and social change, how you do it is often more important than what you do.

DOI: 10.1201/9781003449935-6

6. If the politics don't fly, the hardware never will.
7. Four questions, the Four Who's, need to be answered as a self-consistent set if a system is to succeed economically; namely, who benefits? who pays? and, as appropriate, who loses?
8. Relationships among the elements are what give systems their added value.
9. Sometimes it is necessary to expand the concept to simplify the problem.
10. The greatest leverage in architecting is at the interfaces.

In the architecting process, the vision or method by heuristics complements and reinforces the quality and possibility of better management of proposed complex architectures. This characteristic or method extracts from the professional a more flexible way of expressing or transmitting his knowledge, that is, his wisdom from experiences accumulated in his professional life.

We use heuristics to evaluate the strength and quality of what is currently offered to stakeholders, facilitate critique during planning, design, and development, and predict the effectiveness of a potential solution.

Heuristics usually take the form of short expressions in natural language. These can be memorable phrases encapsulating rules of thumb, shortcuts, or "words to the wise," giving general guidelines on professional conduct or rules, advice, or guidance on how to act in specific circumstances. Common heuristics don't sum up everything there is to know, but they can act as useful entry points to learning more. This chapter presents an overview of heuristics in general, as well as some of those that specifically support the practice of Systems Engineering.

Using heuristics does not guarantee success under all circumstances, but the usefulness of a heuristic can be maximized if the known extent of its applicability is made clear. At their best, heuristics can act as aids to decision-making, value judgments, and assessments [3].

Heuristics have always played an important part in the history of engineering and shaped its progress, especially before science developed to the point when it could also assist engineers. Systems Engineering is still at a stage at which there is not a sufficiently reliable scientific basis for many of the systems being built, which has triggered a renewed interest in heuristics to fill the gap. This is especially true as the practice of Systems Engineering is extended to provide solutions to inherently complex, unbounded, ill-structured, or "wicked" problems [4].

PROS AND CONS OF HEURISTIC EVALUATION

As other methods in research, there are both pros and cons in the usability inspection method of heuristic evaluation. Let's look at a few of them.

Pros of Heuristic Evaluation

- Heuristics can help highlight potential usability issues early in the design process.
- It is a fast and inexpensive tool compared with other methods involving real users.

Cons of Heuristic Evaluation

- Heuristic evaluation depends on the knowledge and expertise of the evaluators. Training the evaluators or hiring external evaluators might increase the time and money required for conducting the evaluation.
- Heuristic evaluation is based on assumptions about what "good" usability is. As heuristics are based on research, this is often true. However, the evaluations are no substitute for testing with real users. These are, as the name suggests, only guidelines and not rules that are set in stone.

In summary, we have proposed the following heuristics to help overcome and avert problems when developing joint operational architectures [2]:

- Lack of knowledge in the domain equals architecture pain. A readily available network of subject matter experts makes the architecture relevant.
- To have a joint vocabulary, use the joint dictionary. Seek a common understandable vocabulary by referencing joint standards and the joint dictionary.
- When defining a joint process stall, there needs to be a process owner to make firm calls. When establishing an enterprise-wide

operational architecture, there needs to be one boss to overcome irreconcilable differences across stakeholders.

- Abstract too high—the models can lie; abstract too low—one gets lost in the flow. Architect at the level of abstraction that provides the answers sought.
- For the process to rule the show, organizational bias must go. People tend to think "organization" first, not "process," and architecture models should be created independent of the organization.
- Confining the architecture to only one level of war can make the architecture poor. Follow the process and information flows; do not limit context to operational or tactical level if not a necessary constraint.
- Critically examine hollow transfer activities to prevent distribution problems and lack of interoperability. Be critical of hollow transfer activities and ensure they have the appropriate visibility to prevent and address capability gaps.

As systems increase in complexity, the architect's job will continue to be tested. These simple heuristics can help increase interoperability and the gains produced from architectural development in the DoD. One kind of heuristic that needs attention with focus on the architecting process is the level of abstraction.

Level of Abstraction Heuristic

An important point always questioned by architects in the process is always the level of granularity of the analyzes that need to go so that they can lead the architecting process. This item is very difficult to conduct, not only for the architect or the architecture team involved. Chapter 4 tells us about how to deploy functions with some level of granularity.

In general, the correct level of abstraction, in general, is what addresses the issues for decision-making or, for example, satisfies the requirements or needs placed in the project. Therefore, abstract too high—the models can lie. Abstract too low—one gets lost in the flow. Finding the right level of abstraction is critical in ensuring the architecture can be communicated effectively and still be useful for its intended purpose [5].

Another important and relevant discussion about heuristics is the point that we are conducting the architecting process and we need to understand the adversities that may be possible in your choices into the arrangement architecture solution considering the operation context. The modeling tools help the architects a lot to include the operation, strategies, and other one's adversities attributes that we need to consider during the process and address the resilient solutions to design allocation throughout the architectural decisions.

These heuristics consider safety as an attribute that we need to keep attention on the architecting process, some architectural frameworks do not consider this attribute in the framework views explicit. Therefore, the resilient systems architecture needs to consider a lot of heuristics and a lot of attributes developed throughout the lessons learned cases and experience.

Flexibility and efficiency are necessary to think about possible paths for people in different possibilities of their journey (both for those who are having the first contact with the product, and for those who have been using it for a longer time, for example).

These sets of heuristics mention architectural decisions in level of design, these heuristics help us understand the importance of monitoring and closing loops in the architecting process during the system development life cycle.

When you compare Chapter 5 that we use function methods, we have a clear structure to deploy functions and conduct the function analysis in the core of architecting process as shown in Figure 5.2, but with the heuristic method, we need to be able to more sensible and consider a professional experience in the architect position and use a knowledge buffer to provide a stand and strong or resilient solution. There are a lot of things we need to do when in the heuristics method is incorporated into the architecting process. One of than is that, we need a bigger architecting team and some of SMEs (Subject Matter Expert) to support your insight when you need to compare your lessons learned or when you access your long-time memory. This memory is responsible for your association knowledge, therefore, we need to check with a bigger team and SMEs if the insight is sometimes valid.

Here we have a set of heuristics as an attribute that expresses the idea of lessons learned or experience captured during professional life in the architect or engineer life.

Resilience Heuristics

Chapter 3 provided an overview of the process to make a human-made system resilient to adversity. The following heuristics have been useful in creating an architecture to make a system resilient:

Heuristic 1: The system is capable of resisting adversity and its impact and effect. This heuristic is derived from the *absorption* technique and supports the Capacity attribute. This heuristic calls for the system to be designed to withstand all anticipated adversities. A seawall to protect a city from an approaching hurricane is an example.

Heuristic 2: The system resists adversity and its impact and effect to a defined extent. This heuristic defines the margin of safety for a system design. This heuristic supports the Capacity attribute.

Heuristic 3: To be resilient, the system is capable of restructuring itself. This heuristic supports the Flexibility attribute. This heuristic provides the maximum configuration options for a successful system. Total system failure is possible if the right configuration is not found and there is no opportunity for repair. Apollo 13 is an excellent example of this heuristic.

Heuristic 4: To be resilient, a system is capable of avoiding an adversity by detecting the adversity in advance and making corrective action. This heuristic supports the Buffering attribute. This heuristic supports the system that is capable of dealing with approaching adversities that have not yet been encountered. Approaching adversities may damage or destroy the system if not dealt with. Resistance to volcanic ash from a volcano is an example.

Heuristic 5: The system has functional redundancy or distributed capability. This heuristic supports the Tolerance attribute. This heuristic allows for retention of capability with variable magnitude. This heuristic assures all capability is not lost when the system is partially damaged. This heuristic is applicable when the system loses the functionality of part of the system. Sully Sullenberger employed this heuristic when his airplane hit a flock of geese in 2020 in the US Airways flight 1549 case study.

Heuristic 6: To be resilient a system is designed with multiple identical branches. This heuristic supports the Capacity attribute. This heuristic is based on the technique of physical redundancy. If this heuristic is not used, dependency on one branch will leave the system highly vulnerable [6].

Heuristic 7: The system is designed with as many independent nodes as possible. This heuristic supports the Tolerance attribute. This heuristic is known as the distributed capability principle. Loss of one node due to an adversity will leave many functioning nodes. A windmill fan is an example of this heuristic in action.

Heuristic 8: The system is designed for alternative means of recovery through human intervention when possible. This heuristic supports the Adaptability attribute. This heuristic provides the flexibility of enabling a human to take actions in response, especially to very unusual situations. Neil Armstrong employed this heuristic when he was landing the Apollo 11 capsule on the moon.

Heuristic 9: The system is designed to allow the human operator to be in command. This heuristic supports the Adaptability attribute.

Humans can maintain the vigilance required to take timely and opposite action. This heuristic is essential because assured recovery is essential.

Heuristic 10: System provide an opportunity for repair or restoration after damage. This heuristic supports the Cohesion attribute. A partially damaged system should be capable of returning, or being returned, to a prior level of capability following an encounter with an adversity. The Hubble satellite is an example.

Heuristic 11: The system includes sensor-based detection of adversity and appropriate response action. This heuristic supports the Buffering attribute. This is a key heuristic in detecting and encountering adversity.

Heuristic 12: Resilience techniques are built in rather than added on the fly. This heuristic supports the Adaptability attribute. This heuristic assures that adversity is addressed throughout the time in transit.

REFERENCES

[1] E. Rechtin, *Systems Architecting: Creating and Building Complex Systems.* Englewood Cliffs, NJ: CRC Press, 1991.

[2] M. W. Maier and E. Rechtin, *The Art of Systems Architecting*, 3rd ed. Boca Raton, FL: CRC Press, 2009. 1991.

[3] B. E. Board, "The Guide to the Systems Engineering Body of Knowledge (SEBoK)." https://sebokwiki.org/wiki/Guide_to_the_Systems_Engineering_Body_of_Knowledge_(SEBoK) (accessed 28 August 2023).

[4] C. W. Churchman, "Wicked Problems," *Management Science*, 1967.

[5] D. AT&L, "Heuristics for Joint Architecting," 2006.

[6] N. Leveson, *Safeware: System Safety and Computers.* Reading, MA: Addison Wesley, 1995, p. 434.

Cybernetics

7

According to Hitchins [1, 2, p. 19], cybernetics is all about controlling entities that are either machines or organisms. He also states that organisms must perform certain basic functions in order to survive. It must overcome shocks and traumas and return to a state of normality. Wiener [3] defines cybernetics as "control and communication in the animal and the machine." The discussion below will cover both mechanistic control and organismic control. This recovery is called resilience as explained in Chapter 5 on resilience.

AN ORGANISMIC ENTITY

Within an organismic entity, there must be rules to encourage interaction and cooperation and discourage conflict. There will be mechanisms that deal with members of the organ that do not observe the rules. There will be sanctions on one hand and mechanisms, such as schools that train the members to conform.

Organismic entities have a life cycle on their own with a beginning, a development, maturity, and eventually a collapse. The organismic entity is itself a system made up of subsystems that perform in some way on behalf of the whole system. In the human body, for example, there needs to be an immune system to suppress pathogens. Organismic entities have much in common with human systems as described in Chapter 1.

A MECHANISTIC ENTITY

Similarly, some entities perform as machines. For example, the human heart is a pump, the brain is a processor, the eyes are a camera, the kidneys are a

 DOI: 10.1201/9781003449935-7

filtration unit and so on. For some observers, for example Bertalanffy [4] life is a machine. Life, the machine continuously exchanges components, contains fuel, and maintains itself. It maintains itself in an ordered flow of processes. Hence, as it is seen, the mechanistic view of cybernetics has much in common with Systems Engineering as described in Chapter 5. Mechanistic entities have much in common with physical systems as described in Chapter 1.

MECHANISTIC CONTROL

Cybernetics offers a view of control that is neither fully organismic or mechanistic. The basic concept is that an input–output signal is fed back and compared to an input signal. This concept can be used for both mechanistic entities and organismic entities as described above.

ORGANISMIC CONTROL

Organismic control is different from mechanistic control in that two control influences are formed that balance the system between two opposing extremes. Sweat in the human body is one example.

For example, if the organ is a human body, then any attempt to keep the temperature between a maximum and a minimum temperature will qualify as organismic control.

SELF-DETERMINATION THEORY (SDT)

This is another theory related to organismic control. According to the Organismic Dialectical approach of SDT [5],

> Self-Determination Theory (SDT) is built around a standpoint known as the organismic dialectical approach. Under this notion, it's assumed active organisms, such as people, have an innate desire to grow, gain new perspectives, and master challenges around them. According to the *organismic* dialectical perspective, meeting experiences and challenges helps you create a sense

of self and encourages your maximum human potential. This process is not automatic, however. The approach notes that the natural drive for growth requires proper social support to thrive. This would account for some differences between individuals.

CHATELIER'S PRINCIPLE

This principle articulated by French chemist Henry-Louis Chatelier applies directly to the idea of control as realized by both the organismic and mechanistic processes reflected in cybernetic processes. According to this principle, "when a constraint is applied to a system in equilibrium then, so far as it can, the system will adjust itself to oppose the constraint."

Le Chatelier showed that

> if a dynamic equilibrium is disturbed by changing the conditions, the position of equilibrium shifts to counteract the change to reestablish an equilibrium. If a chemical reaction is at equilibrium and experiences a change in pressure, temperature, or concentration of products or reactants, the equilibrium shifts in the opposite direction to offset the change.

In the field of chemistry, for example, Le Chatelier showed, for example, that ethanol can be produced by the direct hydration of ethene. The analysis looks at the effect of proportions, temperature, pressure, and catalyst on the composition of the equilibrium mixture and the rate of the reaction.

FEEDBACK LOOPS

Cybernetic phenomena are reflected in feedback loops as described by Meadows [6]. According to Meadows,

> a feedback loop is a closed chain of causal connection from a stock, through a set of decisions or rules or physical laws or actions that are dependent on the level of the stock and back again through a flow to change the stock.

Meadows also shows that feedback loops may be balanced. She says, "balancing feedback loops or goal seeking structures in systems and both sources of stability and sources of resistance to change."

Meadows also discusses runaway loops. She says that runaway loops "are found wherever a system element has the ability to reproduce itself or grow as a constant fraction of itself."

MANAGEMENT CYBERNETICS

Management Cybernetics is the area of cybernetics that concerns itself with management and organizations. The concept was first introduced by Stafford Beer in the 1950s. Mathematician Norbert Wiener defines Management Cybernetics as the science that involves the management of biological and mechanical systems using feedback [7].

This topic provides managers with a clear picture of the approach to complex situations and processes. This is about how organizations can respond to complexity in an effective and adequate manner, which will create a stable, sustainable, and flexible enterprise.

MANAGEMENT CYBERNETICS IN ORGANIZATION

It is important that managers observe the following functions to management Cybernetics:

- It is oriented toward the effect of processes within an organization and looks at the cohesive part of these processes.
- It is oriented toward the collection and application of existing knowledge and the gathering and application of new knowledge.
- In complex situations it helps to create the right approach. It develops the responsibility borne by the individual because of which other people become aware of their own responsibilities.
- It is an approach that is understandable for each employee, and it is geared toward the long term.
- This approach has been practiced by everyone at some point, regardless of whether they are aware of this or not.

CIRCULAR SYSTEM

Cybernetic processes have been described as circular systems; when one process ends, another process starts again from the beginning. This is why processes become visible and why they can be described clearly, because of which they can be used within an organization.

> Assignment: Identify a system that employs feedback as part of its operation. Describe how feedback aids in the functionality of that system.

REFERENCES

[1] P. Checkland, *Systems Thinking, Systems Practice.* New York: John Wiley & Sons, 1999.

[2] D. Hitchins, *Systems Engineering: A 21st Century Systems Methodology (Wiley Series in Systems Engineering and Management).* Hoboken, NJ: John Wiley & Sons, 2007.

[3] N. Wiener. "Cybernetics." https://en.wikipedia.org/wiki/Cybernetics (accessed 8 August 2023).

[4] L. V. Bertalanffy, *General Systems Theory: Foundation, Development, Applications,* Revised ed. New York: George Baziller, 1968.

[5] SDT. "Self-Determination Theory (SDT)." https://en.wikipedia.org/wiki/Self-determination_theory (accessed 8 August 2023).

[6] D. H. Meadows and D. Wright, Ed. *Thinking in Systems.* White River Junction, Vermont: Chelsea Green Publishing, 2008.

[7] V. Van Vliet (2011). *Management Cybernetics by Stafford Beer.*

System Discipline Interactions

8

One of the main purposes of this book was to describe five different disciplines for which systems theory is a core discipline and to show how these disciplines have various characteristics in common and how they interact with each other. The five disciplines are: Systems Theory, Systems Science, Systems Engineering, Systems Architecting, and Systems Thinking. The purpose of this chapter is to present various terms and concepts discussed throughout the book and show where they appear. The conclusion is that the book itself is an abstract system, as described below, and that the terms and concepts are the interactions between the various parts.

HIERARCHY

This is a feature of all systems, both real and abstract, and therefore is the subject of all chapters in this book. It was first introduced in Chapter 1, but it was pointed out that hierarchy itself is not necessarily physical. It is therefore a human-perceived feature that can be realized in many ways, for example, as an abstract plan to design a system. Hierarchies of functions are also a way for hierarchy to manifest itself.

HOLISM

This is another concept introduced by Smuts [1] as a way to show that the entire system is larger than the sum of its parts. This is done by including the

DOI: 10.1201/9781003449935-8

emergent properties (see also Chapter 2) into the description of the whole system. Smuts also points out that the holism view also includes all of the activities of the parts. Jackson [2] also points out that holism includes the functions of the parts. This concept is described in Chapter 2.

BOUNDARIES

Like hierarchy and holism, all systems have boundaries, and therefore this concept applies to all chapters in this book. As Chapter 2 states, some practitioners have difficulty in identifying the boundaries. Chapter 2 provides the following rule: A boundary is the point for which the system no longer exhibits emergent properties. This rule alone ties this concept to the concept of emergence, which is explained in Chapter 2.

PARTS

The concept of parts, also called elements, appears in the definition of system in Chapter 1. Therefore, this concept applies to all chapters in this book. This concept comes into focus in Chapter 2 in which the concept of holism holds that a system is larger than the sum of its [physical] parts.

SYSTEM

The concept of the system is the focus of the entire book; hence, this concept applies to all chapters. The concept of system implies that a system must have at least two parts; otherwise, there cannot be interactions and hence no emergent properties. This concept is defined in Chapter 1.

This concept opens the question of whether there is such a thing as a non-system. As implied in the definition of system, a non-system is an entity that has either (1) only one part, (2) no interactions between parts, or (3) no emergent properties. A single-part entity will meet all three criteria.

FUNCTION

Chapter 1 states that a basic characteristic is its purpose or function. In addition, Chapter 4 shows that functions are entities that show the need for specific system elements. In addition, Chapter 5 shows that functions have two primary reasons for being:

- First, functions are necessary to describe the need for subsystems within the host system.
- Second, functions are necessary to describe the purpose of the host system without which there would be no system.

SUBSYSTEMS

This term is most commonly used in the context of physical human-made systems. Therefore, it is a topic relative to the discipline of systems engineering discussed in Chapter 3. Nevertheless, this concept does not conflict with other discussions in Chapter 4. The concept of subsystems implies a hierarchical arrangement of parts and subsystems. Therefore, it interacts with the concept of hierarchy described in Chapter 2 and above.

HOLON

This is a more limited concept and has its beginning in Chapter 1 regarding natural systems. It is observed that certain systems, primarily natural, have a repeated pattern of subsystems within the concept of hierarchy described in Chapter 2. This occurs when a certain particle, for example, a seed, has within it the beginnings of a larger entity, for example, an apple tree. The apple tree also has seeds within it, which are the beginnings of future trees. So, each level is a holon of another human-made system and is therefore applicable to systems engineering as described in Chapter 4.

EMERGENCE

This is a central idea behind systems science and therefore behind systems theory. It is therefore introduced in Chapter 2. However, it is implicit in Chapter 1. In addition, it applies to all chapters in this book. Emergence is a property of a system that cannot be determined by the examination of a single part.

ARTIFACTS

According to Hitchins [3], artifacts are parts of a system that do not include humans. As an example, an aircraft without a pilot is an artifact or a collection of artifacts. Therefore, this concept is only related to systems engineering as discussed in Chapter 5.

ABSTRACTIONS

Abstractions are entities that are mental and have no physical manifestation. The term conceptual is also often used for abstract entities including systems. Common examples of abstract systems include poems, symphonies, and novels. Of course, when we mention these examples, we are not including the medium of expression, for example, a manuscript. In industry, abstract systems include plans. In short, an abstract system is a system that exists in the mind of the creator or author. Hence, this book is an abstract system or better put, an abstract meta-system as indicated in the title.

In this chapter, hierarchies are abstract entities since they have no physical presence and only exist in the mind of the designer or observer. This chapter as a whole is an abstract system, and the chapters themselves are the parts. This section completes the systemic description since it describes the interactions between the parts, which are the chapters.

COMPLEXITY

Complexity is a characteristic of a system that contains the following properties, according to [4]:

- Different kinds of parts,
- Each kind of part,
- Kinds of interactions,
- Part aggregation, and
- Behaviors.

Watson et al. [5] elaborate on these characteristics. In summary, all systems are not complex. Complexity adds a degree of uncertainty that makes them difficult to analyze. Yet, they are still systems as described in Chapter 1 and are therefore a topic of interest for this book. This concept is described in Chapter 1.

INTERACTIVITY

Chapter 1 states that this concept is a basic characteristic of all systems. In addition, all systems involve interactivity; therefore, this concept is applicable to all chapters. The interactivity between Chapter 5 (Functions) and Chapter 6 (Heuristics) lies in the fact that the use of functions is itself a heuristic; Maier and Rechtin [6, p. 403] state, for example, that "except for good and sufficient reasons, functional and physical structuring should match."

BEHAVIOR

Chapter 1 identifies behavior as a basic cause of complexity.

DYNAMICS

Chapter 1 identifies dynamics as a basic cause of complexity.

SYSTEMS OF SYSTEMS

This concept is discussed in Chapter 1.

COMPLICATED

Chapter 1 describes a complicated system as one for which the interactions are static and predictable.

HUMAN-MADE

Chapter 1 identifies human-made systems as a basic type of system.

NATURAL SYSTEMS

Chapter 1 describes human-made systems as systems not designed by humans.

CYBERNETICS

This is a term used to describe characteristics of a system that is seeking a viable state in between two extreme conditions. To do this, most systems require a feedback loop in their architecture.

REFERENCES

[1] J. C. Smuts, *Holism and Evolution*. Gouldsboro, ME: Gestalt Journal Press, 1926.

[2] S. Jackson, "Functional Holism of Systems," *System Research and Behavioral Science*, 2022.

[3] D. Hitchins, *Systems Engineering: A 21st Century Systems Methodology (Wiley Series in Systems Engineering and Management)*. Hoboken, NJ: John Wiley & Sons, 2007.

[4] G. E. Mobus and Michael C. Kalton, *Principles of Systems Science*. New York: Springer, 2015.

[5] M. Watson, R. Anway, D. McKinney, L. A. Rosser, and J. MacCarthy, "Appreciative Methods Applied to the Assessment of Complex Systems," presented at the *INCOSE International Symposium*, Orlando, FL, 2019.

[6] M. W. Maier and E. Rechtin, *The Art of Systems Architecting*, 3rd ed. Boca Raton, FL: CRC Press, 2009. 1991.

Systems Thinking

9

INTRODUCTION

The essence of this chapter is that anyone should know that the universe is full of systems and that everyone should know what those systems are, and that they should be able to recognize the systems in which they live or exist.

Primary sources for this topic are [1–3]. The two most obvious answers to this question are (1) the universe and (2) humanity. We will expand on this idea in this chapter.

First, we need to remind ourselves of what a system is. There are four answers to this question. They are: (1) a system must consist of many parts, (2) the parts must interact, and (3) the collection of parts (the system) should exhibit emergence, which is a property not identifiable by individual parts. The fourth quality is that the system should have a boundary. This quality was elaborated in Chapter 1. The most obvious system is the universe. This system contains all the physical systems you can imagine. The second most obvious system is humanity. This system contains all humans of which you are one. This system also contains abstract systems, such as government. The physical government can be constructed from the abstract system.

The Five Paths of Systems Thinking

According to Hoverstadt [3], there are five paths to systems thinking.

1. The first path is to identify the emergent property of any system. According to Hoverstadt [3], this is not as easy as it may seem. As a start, Hoverstadt recommends looking at "the pattern of boundaries, relationships, and dynamics, the effects they have on one

DOI: 10.1201/9781003449935-9

another, and what that creates." Even with all this advice, it still may be very difficult. Is it possible, for example, to predict the taste of salt? Probably not. This path has a strong relation to section on Emergence in Chapter 2 of this book, even though achieving this path may be difficult.

2. The second path is identifying the system of the emergent. This path is much easier than the first path. This is because most people are familiar with the emergent aspects of common systems. For example, if you hear a bark, you immediately know the system is a dog. Hence, this path also has a strong relation to section on Emergence in Chapter 2 of this book.

3. The third path is recognizing how the emergent property changes the system. Hoverstadt [3] uses as an example the way that corporations change the behavior of people. This may be a more subtle way that emergence changes systems, but it is more evident on closer inspection. This path also shows the strong relation between this path and section on Emergence in Chapter 2 of this book.

4. The fourth path is recognizing how emergence changes over time. Hoverstadt [3] points out that many systems become more stable over time. Both the weather and the stock market are common examples. These also show the strong relation between this chapter and section on Emergence in Chapter 2 of this book.

5. The fifth path is to recognize how events can be unpredictable, the "outrageous fortune" aspect. These include mass migration and global warming. Whatever the phenomena, this path further reinforces the relation between this path and section on Emergence in Chapter 2 of this book.

The Meadows View of Systems Thinking

One of the most intelligent and cogent summaries of Systems Thinking is Meadows [1]. Probably the best way to summarize her observations is just to quote them:

1. First, she quotes Persig, author of the best-selling book *Zen and the Art of Motorcycle Maintenance*, as follows: "There is so much talk about systems and so little understanding."

2. She has a number of observations about the misunderstanding of the importance of systems:
   - Political leaders don't cause recessions. They are inherent in the structure of a market economy.

- Competitors rarely cause a company to lose market share. The losing company creates its own losses through its own policies.
- Oil-exporting nations are not solely responsible for the rise in oil prices. The cause is in the policies that made them vulnerable to supply interruptions.
- The flu virus does not attack you; you set up the conditions that let it flourish within you.
- Drug addiction is not the fault of an individual; the fault lies in a larger set of influences.

3. She says that Systems Thinking allows one to see interconnections. (See section on Interfaces in Chapter 5.)
4. She says that systems "aren't just any old collection of things. A system is an interconnected set of elements organized in such a way that achieves something. A system must consist of three things: elements, interconnections, and a *function* or *purpose*."
5. She says that "systems are resilient and many of them are evolutionary." She is probably talking about inherent resilience here, as opposed to the designed-in resilience discussed in Chapter 5.
6. She says, "Once you start listing the elements of a system, there is almost no end to the process. You can divide elements into sub-elements and then sub-sub-elements. Pretty soon you lose sight of the system." She says, "It is easier to learn about a system's elements than its interconnections."
7. She says that an important function of almost every system is to ensure its perpetuation.
8. She says that "the least obvious part of a system, its function or purpose is often the most crucial determinant of the system's behavior." "Interconnections are also critically important."
9. Meadows also discusses the need to understand feedback loops in Systems Thinking as follows: Feedback loop is a closed chain of causal connection from a stock, through a set of decisions or rules or physical laws or actions that are dependent on the level of the stock and back again through a flow to change the stock.

Although the above observations may seem obvious, they can also be considered wise.

REFERENCES

[1] D. H. Meadows and D. Wright, Ed. *Thinking in Systems*. White River Junction Vermont: Chelsehat Green Publishing, 2008.

[2] P. Checkland, *Systems Thinking, Systems Practice*. New York: John Wiley & Sons, 1999.

[3] P. Hoverstadt, *The Grammar of Systems*. Las Vegas, NV: SCiO Publications, 2022.

Final Thoughts

10

THE VALUE OF SYSTEMS THINKING

If the goal of the reader is to become a better Systems Thinker, he or she should start by reading Chapter 6, Systems Thinking. This chapter focuses on the thoughts of two writers, Hoverstadt [1] and Meadows [2]. Hoverstadt focuses on the understanding of emergence and how it is realized. This is one of the most difficult aspects of Systems Thinking.

Meadows, on the other hand, focuses on many difficult aspects of Systems Thinking. These include understanding what a system is and the importance of purpose; Hoverstadt, on the other hand, also emphasizes the importance of understanding emergence and the possible unpredictability of it. She also discusses the common misunderstandings of systems. She emphasizes the importance of understanding interconnections and resilience, the elements of a system, and that a function of systems is to perpetuate themselves.

As for the book itself, one of the primary goals was to explain how the various disciplines that comprise Systems Theory interact and complement each other. This is an aspect that is not often explained. There are many aspects that many of these disciplines have in common. These include hierarchy, holism, boundaries, and interactions. Many of these emanate from the discipline of Systems Science, one of the primary drivers of Systems Theory.

As a final thought, it is not far-fetched to think of a string quartet as a metaphor for Systems Theory in which each instrument complements the other instruments. To take that metaphor even further, think of some of the disciplines as natural systems and others as human-made. This idea makes the interconnections factor even more awesome.

DOI: 10.1201/9781003449935-10

THE FUTURE OF SYSTEMS

So, where do we go from here? The science publication New Scientist [3] provides an overview of the promises and threats of the popular field of computing commonly called artificial intelligence (AI). This field can be seen as relevant to systems because it focuses largely on human decision makers, and we saw in Chapter 1 that humans are systems.

The aspect that makes this technology so human is that it consists of neural networks (like humans) and types of neural networks called transformers. In this book, we have discussed many system functions, such as making decisions. So, if a machine can imitate human brains and make decisions, why do we need humans? Experts agree that we are not at a level of maturity yet, but it is almost certain that someone is working on it.

There are many areas of potential benefit. For example, the elimination of war and hunger would be ideal candidates. However, it is not known whether the level of maturity is great enough to tackle these issues or not.

But first, we need to look at a few areas of progress. One of these is facial recognition. This is an area where the technology is encroaching on human talents. Another area of progress is recognizing handwriting. This is also a human talent. Then there is medical diagnosis, also known as triage. It has been found that this technology is effective in this area, another human capability. The technology has also been found to be effective in another area of human expertise, namely, the creation of language models. In short, the primary focus of the technology is altering neural networks, so they more closely resemble the web of neurons in our brains. The primary focus of the detractors is the concern that technology may replace us and make decisions we have no control over. There is no indication that technology is headed in this direction, but it is not too soon to begin taking steps to prevent this outcome.

Hence, let's focus our attention on a goal that may or may not be more tractable, namely, the elimination of catastrophes. Elsewhere in this book we have studied some recent catastrophes, all of which were attributed to human error to some degree. In Appendix A, we studied four major catastrophes. These were (1) the Air Blue 202 accident, (2) the *Challenger* accident, (3) the *Columbia* disaster, and (4) the *Concorde* accident. The question to be asked is whether any or all of these accidents could have been prevented if more advanced technology could have been invented to help make the decisions that could have saved lives. If this sounds like a "pipe dream," so be it. On the other hand, the technology seems to be progressing at such a rate that these types of decisions could very well be within the grasp of system designers.

So, what does it take for a computer to emulate a brain? The esteemed MIT Professor Leveson has stated that the two most common causes of system failure are (1) complacency and (2) optimism. Both of these are well-known and undesirable human characteristics. Furthermore, there are no known systems or processes that will eliminate complacency or optimism. So, emulating human decision-making will not be easy. However, many experts in the field say that we are well on the way.

Before we go any further, we need to re-emphasize the preliminary nature of the goals already cited. For example, there is only speculation that technology has the capability of reducing or eliminating warfare. On the other hand, the evidence is very promising for the use of technology to assist in making important decisions.

In the above paragraphs, we covered human decisions very well and concluded the use of technology to make these decisions in the future cannot be ruled out. We cannot, however, place a projected date on these predictions. In addition, we cannot say for certain whether these predictions can even be realized even though the progress seems to be moving forward.

We have already discussed human decision-making and found that progress is being made. But, how about the old-fashioned types of decisions, that is, the technical kind? The launch of a rocket would fall into that category. It is not unreasonable to think that advanced technology could analyze an entire system, such as a spacecraft or a commercial airplane like the Air Blue plane discussed in the next chapter.

Therefore, the next logical question to ask is whether any technological system has the capability of examining any technological system to the component level and determine whether there are any flawed components that could lead to a catastrophe. The obvious example is the O-ring associated with the *Challenger* disaster. At first glance, this prospect would not seem any more difficult than modeling the human brain, which has already been seen to be, at least, on the horizon. Any large system like this may contain thousands, if not millions, of components. Let's not close out this discussion by giving the impression that success is certain. It certainly is not. On the other hand, it is difficult not to be impressed by the progress so far. So, let's just cross our fingers and hope for the best, and don't ignore the detractors. Their cautions must be heeded.

REFERENCES

[1] P. Hoverstadt, *The Grammar of Systems*. Las Vegas, NV: SCiO Publications, 2022.

[2] D. H. Meadows and D. Wright, Eds. *Thinking in Systems*. White River Junction, Vermont: Chelsea Green Publishing, 2008.

[3] M. Sparkes, "Artificial Intelligence: 5 Questions Answered That You Should Know." *New Scientist*, 2023.

Case Studies

This appendix documents four cases designed to illustrate lessons learned from cases in which decisions were made with catastrophic outcomes. The cases are as follows:

- A. Air Blue Flight 202. This case illustrates an example of an accident for which egregious cognitive biases were a primary contributing factor.
- B. *Challenger*. This is a case for which it is shown that inadequate attention to risk was shown.
- C. *Columbia*. This is a case for which authority to launch was not provided by an independent technical authority (ITA).
- D. *Concorde*. This is a case for which complacency was the primary cause.

CASE A

Airblue Flight 202 [1] was a scheduled Pakistani domestic aircraft from Karachi to Islamabad. On July 28, 2010, the Airbus A321-231 airliner serving the flight crashed, killing all 146 passengers and six crew. Both the captain and first officer were fully qualified to fly the aircraft. Following is a partial account of the incident:

Pakistan's Civil Aviation Authority [1] stated that the crash was caused by the captain's incorrect flight-control input. The captain's arrogant demeanor (sic) during the entire flight made him less likely to speak up on the captain's mistakes and effectively ceased any meaningful information about the accident.

The aircraft crashed in the mountains outside the airport of Islamabad, The pilot did not send any emergency signals prior to the crash. The recording of the FDR and CVR revealed that the pilot in control of the aircraft at the time had significantly deviated from the established procedures for an approach to Islamabad. At one important point, the pilot could be heard laughing off the order from the ATC to enter a holding pattern and wait for the weather to improve, saying, "Let him say what he wants to say."

According to the recording, he initially did not feel anxious about his decision to deviate from the standard procedures for an approach to Islamabad.

His arrogant attitude makes this case a perfect example of the Optimism bias.

74

CASE B

Challenger. This case makes a good example for Decision Management because it illustrates the importance of attention to risk. Much of the literature concentrates on the failure of the O-rings. However, the O-rings failed because of the prolonged exposure to low temperatures. The sensitivity to temperature was known to management. Hence, this factor was a result of inattention to temperature.

In her exhaustive book, Vaughn [2] illustrates the climate in which risks are "normalized." That is, it isn't that risks were not known; it is that they were not taken seriously. For this reason, the *risk denial* bias, a human trait, seems to have prevailed even if tacitly.

Then there is the matter of communication. The communication system on the program was such that if a person at the lower levels of the organization hierarchy has a concern, the likelihood of that concern reaching a person of importance was very slim.

If there is a simple message for this case, though, it is to *take risks seriously*. This message is more than a mental attitude. It involves all the procedures involved in taking risks seriously. For example, how are risks communicated? How are they recorded and reviewed? And finally, how are actions identified and acted on? It is assumed that the space program is addressing all these questions and executing the required actions.

CASE C

Columbia. The amazing thing about the *Columbia* accident is that it happened only a few months after the Challenger accident, and all the lessons had not yet been put into action. The Columbia accident was followed by an exhaustive report called *The Columbia Accident Investigation Report* [3].

Before we discuss the report, though, we need to review what the primary causes of accidents are. According to the esteemed MIT Professor Leveson [4], there are two primary causes of accidents: *complacency* and *optimism*. The thing both these causes have in common is that they are highly *judgmental*. That is, no one has a complacency meter to tell when a supposed complacency limit or optimism limit has been exceeded. Furthermore, there does not seem to be a complacency limit or optimism limit identified.

What can be done, though, is to be clear about how to mitigate risks. The report provides a clear answer to this question: It recommends that any future actions should be approved by an ITA before action is taken.

The first part of the ITA is the word *independent*. This word is normally interpreted to mean both organizationally and financially independent. That is, the ITA team or committee must be both organizationally and financially independent of the host organization. It can be from some other branch of the government but not the same branch. Similarly, the host organization must be financially independent. That is, it cannot receive funding from the same branch of government. Similarly, the technical evaluation of the proposed action, for example, a launch approval should be made organizationally and financially independent of the host organization.

CASE D

Concorde. Another well-known disaster was the crash of the British-French supersonic aircraft called the *Concorde*. On July 25, 2000, Air France Flight 4590, a *Concorde* passenger jet on an international charter flight from Paris to New York, crashed shortly after takeoff, killing all 109 people on board and four on the ground. The accident investigation by the French agency revealed that Air France 4590's right front tire on the left main landing gear was destroyed after having run over a strip of metal debris shed by a Continental Airlines DC-10-30, which departed on the runway just five minutes prior. *Concorde* jets went back into service, but a series of minor problems prompted both Air France and British Airways to end Concorde service permanently in October 2003.

But Continental labeled the initial court decision absurd and launched an appeal. A French appeals court has cleared US airline Continental of criminal blame for the July 2000 crash of a *Concorde* jet shortly after takeoff from Paris. The airline's parent firm, United Continental Holdings, said in a statement that it supported the court's decision that Continental was not to blame.

However, since the accident, an independent analysis was conducted [5] which concluded that there were many more factors that could have contributed to the accident. According to the report,

> the plane was operating outside its legally certified limits. When it stood at the end of the runway, ready to roll, it was more than six tons over its approved maximum takeoff weight for the given conditions, with its center of gravity pushed dangerously far to the rear.

The US airline had consistently argued that the *Concorde* caught fire before it hit the metal strip and that they were being used as a scapegoat to protect France's airline industry.

Regarding this dispute, this account remains neutral. However, there are some conclusions to draw: First, rigorous risk management is required under all conditions, whether the system is commercial or government managed. Second, a risk mitigation approach is a must.

REFERENCES

[1] Pakistan CAA, "Investigation Report – Air Blue Flight ABQ202," Pakistan Civil Aviation Authority, Karachi, March 7, 2011.

[2] D. Vaughn, *The Challenger Launch Decision: Risky Technology, Culture, and Deviance at NASA*. Chicago, IL: University of Chicago Press, 1997. 1996.

[3] NASA, "Columbia Accident Investigation Report," National Aeronautics and Space Administration (NASA), Washington, DC, August 2003, vol. *1*.

[4] N. Leveson, *Safeware: System Safety and Computers*. Reading, MA: Addison Wesley, 1995, p. 434.

[5] D. Rose, "Doomed: The Real Story of Flight 4590," in *Guardian*, London, 2001.

Additional Characteristics of Systems

In addition to the essential characteristics in Chapter 1, the Fellows noted the following characteristics that systems have in common:

1. Systems have *structure* – comprising parts, and relationships between them.
2. Systems exhibit *holism, whole, togetherness, unity,* or *systemness* – the extent to which something is a system.
3. Systems exhibit *emergence* – the property, function, or phenomenon that can be attributed only to the system as a whole and not to any of its components acting alone.
4. Systems occur at multiple levels of organisation (e.g. cell, organ, organism in living systems), with new types of emergence "emerging" at each level or organisation.
5. Systems exist within their *environment* – with which they interact, and exchange material, energy, and information.
6. These is a *goal, purpose, mission,* or *objective* that the system is expected to achieve or for which it is designed. We argue below that this only applies to artificial, man-made systems.

Index

Pages in *italics* refer to figures and pages in **bold** refer to tables.